*Study and Solutions Guide for*

# CALCULUS

SIXTH EDITION

Larson / Hostetler / Edwards

## Volume II

David E. Heyd
The Pennsylvania State University
The Behrend College

**Houghton Mifflin Company** Boston New York

Editor in Chief, Mathematics: Charles Hartford
Managing Editor: Cathy Cantin
Senior Associate Editor: Maureen Brooks
Associate Editor: Michael Richards
Assistant Editor: Carolyn Johnson
Supervising Editor: Karen Carter
Art Supervisor: Gary Crespo
Marketing Manager: Sara Whittern
Associate Marketing Manager: Ros Kane
Marketing Assistant: Carrie Lipscomb
Design: Henry Rachlin
Composition and Art: Meridian Creative Group



Printed in the U.S.A.

ISBN: 0-395-88768-2

456789-B-01

# Preface

This *Study and Solutions Guide* is designed as a supplement to *Calculus,* Sixth Edition, by Roland E. Larson, Robert P. Hostetler, and Bruce H. Edwards. All references to chapters, theorems, and exercises relate to the main text. Although this supplement is not a substitute for good study habits, it can be valuable when incorporated into a well-planned course of study. The following suggestions may assist you in the use of the text, your lecture notes, and this *Guide*.

- *Read the section in the text for general content before class.* You will be surprised at how much more you will absorb from the lecture if you are aware of the objectives of the section and the types of problems that will be solved. If you are familiar with the topic, you will understand more of the lecture, and you will be able to take fewer (and better) notes.
- *As soon after class as possible, work problems from the exercise set.* The exercise sets in the text are divided into groups of similar problems and are presented in approximately the same order as the section topics. Try to get an overall picture of the various types of problems in the set. As you work your way through the exercise set, reread your class notes and the portion of the section that covers each type of problem. Pay particular attention to the solved examples.
- *Learning calculus takes practice.* You cannot learn calculus merely by reading, any more than you can learn to play the piano or to bowl merely by reading. Only after you have practiced the techniques of a section and have discovered your weak points can you make good use of the supplementary solutions in this *Guide.*
- *Technology.* Graphing utilities and symbolic algebra systems are now readily available. The computer and calculator are merely tools. Your ability to use these tools effectively requires that you continually sharpen your problem-solving skills and your understanding of fundamental mathematical principles.

During many years of teaching I have found that good study habits are essential for success in mathematics. My students have found the following additional suggestions to be helpful in making the best use of their time.

- *Write neatly in pencil.* A notebook filled with unorganized scribbling has little value.
- *Work at a deliberate and methodical pace, without skipping steps.* When you hurry through a problem you are more apt to make careless arithmetic or algebraic errors that, in the long run, waste time.
- *Keep up with the work.* This suggestion is crucial because calculus is a very structured topic. If you cannot do the problems in one section, you are not likely to be able to do the problems in the next. The night before a quiz or test is not the time to start working problems. In some instances cramming may help you pass an examination, but it is an inferior way to learn and retain essential concepts.
- After working some of the assigned exercises with access to the examples and answers, *try at least one of each type of exercise with the book closed.* This will increase your confidence on quizzes and tests.
- Do not be overly concerned with finding the most efficient way to solve a problem. *Your first goal is to find one way that works.* Short cuts and clever methods come later.
- If you have trouble with the algebra of calculus, *refer to the algebra review at the beginning of this guide.*

I wish to acknowledge several people whose help and encouragement were invaluable in the production of this *Guide*. First, I am grateful to Roland E. Larson, Robert P. Hostetler, and Bruce H. Edwards for the privilege of working with them on the main text. I also wish to thank the staffs at Larson Texts, Inc. and Houghton Mifflin Company. I am also grateful to my wife, Jean, and our children, Ed, Ruth, and Andy, for their love and support.

David E. Heyd

# Contents

# CHAPTER 10
# Vectors and the Geometry of Space

# CHAPTER 10
# Vectors and the Geometry of Space

## Section 10.1 Vectors in the Plane

### Solutions to Selected Odd-Numbered Exercises

**5.** (a) The sketch of the directed line is shown in the figure.

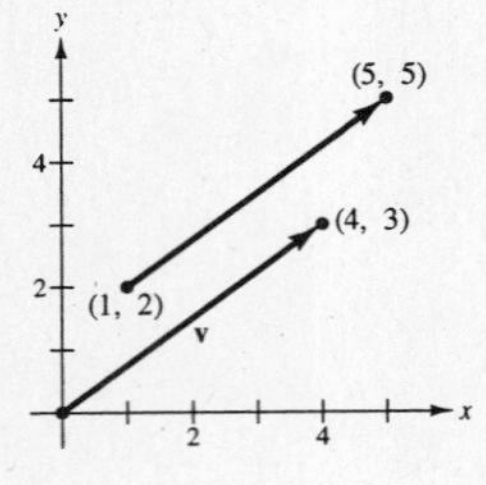


(b) We let $P = (1, 2) = (p_1, p_2)$ and $Q = (5, 5) = (q_1, q_2)$. Then the components of $\mathbf{v} = \langle v_1, v_2 \rangle$ are given by

$$v_1 = q_1 - p_1 = 5 - 1 = 4$$

$$v_2 = q_2 - p_2 = 5 - 2 = 3.$$

Thus, $\mathbf{v} = \langle 4, 3 \rangle$.

(c) The sketch of the vector $\mathbf{v}$ is shown in the figure.

**21.** Since $\mathbf{u} = 2\mathbf{i} - \mathbf{j}$ and $2\mathbf{w} = 2(\mathbf{i} + 2\mathbf{j}) = 2\mathbf{i} + 4\mathbf{j}$, we have

$$\begin{aligned}\mathbf{v} &= \mathbf{u} + 2\mathbf{w}\\ &= (2\mathbf{i} - \mathbf{j}) + (2\mathbf{i} + 4\mathbf{j})\\ &= (2 + 2)\mathbf{i} + (4 - 1)\mathbf{j} = 4\mathbf{i} + 3\mathbf{j} = \langle 4, 3 \rangle.\end{aligned}$$

Geometrically, this sum is illustrated in the figure.

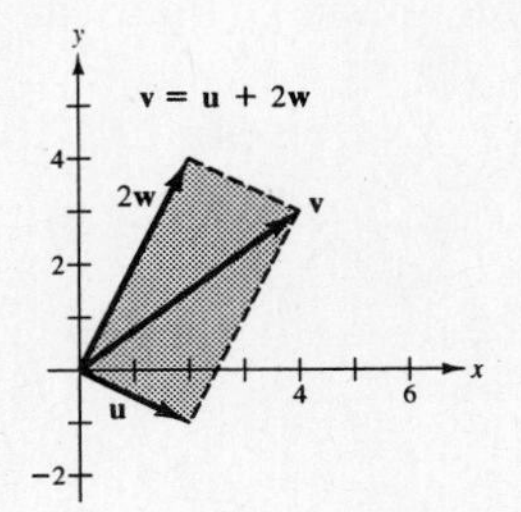


**25.** $\mathbf{v} = a\mathbf{u} + b\mathbf{w}$

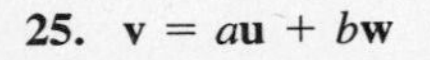

$$3\mathbf{i} = a(\mathbf{i} + 2\mathbf{j}) + b(\mathbf{i} - \mathbf{j}) = (a + b)\mathbf{i} + (2a - b)\mathbf{j}$$

Equating coefficients yields

$$a + b = 3 \text{ and } 2a - b = 0.$$

Solving the system of equations simultaneously we obtain $a = 1$ and $b = 2$.

**33.** The magnitude (length) of the vector $\mathbf{v} = v_1\mathbf{i} + v_2\mathbf{j}$ is

$$\|\mathbf{v}\| = \sqrt{{v_1}^2 + {v_2}^2}.$$

The magnitude of the vector $\mathbf{v} = 6\mathbf{i} - 5\mathbf{j}$ is

$$\|\mathbf{v}\| = \sqrt{6^2 + (-5)^2} = \sqrt{61} \approx 7.81.$$

**41.** $\|\mathbf{u}\| = \sqrt{2^2 + 1^2} = \sqrt{5}$

$\|\mathbf{v}\| = \sqrt{5^2 + 4^2} = \sqrt{41}$

Since $\mathbf{u} + \mathbf{v} = \langle 2, 1 \rangle + \langle 5, 4 \rangle = \langle 7, 5 \rangle$, we have

$$\|\mathbf{u} + \mathbf{v}\| = \sqrt{7^2 + 5^2} = \sqrt{74}.$$

Therefore,

$$\begin{aligned}\|\mathbf{u} + \mathbf{v}\| = \sqrt{74} &\approx 8.602 \le 2.236 + 6.403\\ &\approx \sqrt{5} + \sqrt{41} = \|\mathbf{u}\| + \|\mathbf{v}\|.\end{aligned}$$

**43.** $\mathbf{v} =$ (magnitude of $\mathbf{v}$)(unit vector in the direction of $\mathbf{v}$)

$$\begin{aligned}&= 4\left(\frac{\mathbf{u}}{\|\mathbf{u}\|}\right)\\ &= 4\left(\frac{\mathbf{u}}{\sqrt{2}}\right) = 2\sqrt{2}\mathbf{u} = \left\langle 2\sqrt{2}, 2\sqrt{2} \right\rangle\end{aligned}$$

**47.** (a) At (1, 1) the slope of the tangent line is $f'(1) = 3$. Therefore a vector $\mathbf{v} = x\mathbf{i} + y\mathbf{j}$ parallel to the tangent line must have a slope of 3 (see figure) and $y = 3x$. Since $\mathbf{v}$ is a unit vector, $x^2 + y^2 = 1$. Thus,

$$x^2 + (3x)^2 = 1$$
$$10x^2 = 1$$
$$x = \pm\frac{1}{\sqrt{10}} \text{ and } y = \pm\frac{3}{\sqrt{10}}.$$

Finally, we conclude that $\mathbf{v} = \frac{1}{\sqrt{10}}\mathbf{i} + \frac{3}{\sqrt{10}}\mathbf{j} = \left\langle \frac{1}{\sqrt{10}}, \frac{3}{\sqrt{10}} \right\rangle$ or $\mathbf{v} = -\frac{1}{\sqrt{10}}\mathbf{i} - \frac{3}{\sqrt{10}}\mathbf{j} = \left\langle -\frac{1}{\sqrt{10}}, -\frac{3}{\sqrt{10}} \right\rangle$.

(b) Similarly, a vector $\mathbf{v} = x\mathbf{i} + y\mathbf{j}$ normal to the tangent line must have a slope of $-\frac{1}{3}$ (see figure), and $x = -3y$. Since $\mathbf{v}$ is a unit vector, $x^2 + y^2 = 1$. Thus,

$$(-3y)^2 + y^2 = 1$$
$$10y^2 = 1$$
$$y = \pm\frac{1}{\sqrt{10}} \text{ and } x = \pm\frac{3}{\sqrt{10}}.$$

Finally, we conclude that $\mathbf{v} = \frac{3}{\sqrt{10}}\mathbf{i} - \frac{1}{\sqrt{10}}\mathbf{j} = \left\langle \frac{3}{\sqrt{10}}, -\frac{1}{\sqrt{10}} \right\rangle$ or $\mathbf{v} = \frac{-3}{\sqrt{10}}\mathbf{i} + \frac{1}{\sqrt{10}}\mathbf{j} = \left\langle -\frac{3}{\sqrt{10}}, \frac{1}{\sqrt{10}} \right\rangle$.

**53.** Begin by finding a unit vector $\mathbf{u}$ making an angle of 150° with the positive $x$-axis. Since $\mathbf{u}$ is a unit vector, consider it as the radius of a unit circle as shown in the figure. Therefore, $x = \cos\theta$ and $y = \sin\theta$, and the component form for $\mathbf{u}$ is

$$\mathbf{u} = x\mathbf{i} + y\mathbf{j} = (\cos 150°)\mathbf{i} + (\sin 150°)\mathbf{j} = \left\langle -\frac{\sqrt{3}}{2}, \frac{1}{2} \right\rangle.$$

Since $\mathbf{v} = 2\mathbf{u}$, we have $\mathbf{v} = \langle -\sqrt{3}, 1 \rangle$.

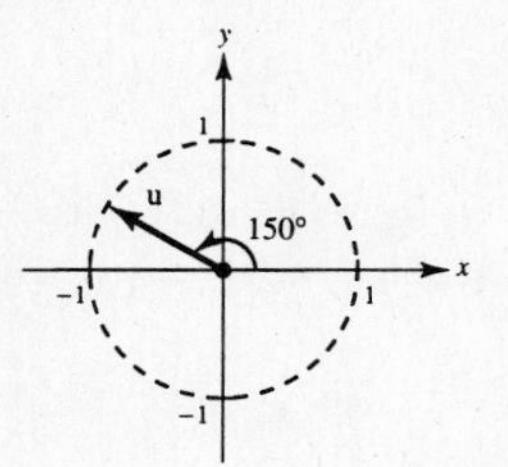


**59.** For any nonzero vector $\mathbf{w}$ making an angle $\theta$ with the positive $x$-axis can be written as

$$\mathbf{w} = \|\mathbf{w}\|(\cos\theta\,\mathbf{i} + \sin\theta\,\mathbf{j}).$$

Since $\|\mathbf{u}\| = 1$ and $\theta = 45°$,

$$\mathbf{u} = 1(\cos 45°\,\mathbf{i} + \sin 45°\,\mathbf{j}) = \frac{\sqrt{2}}{2}\mathbf{i} + \frac{\sqrt{2}}{2}\mathbf{j}.$$

Since $\|\mathbf{u} + \mathbf{v}\| = \sqrt{2}$ and $\theta = 90°$,

$$\mathbf{u} + \mathbf{v} = \sqrt{2}\,(\cos 90°\mathbf{i} + \sin 90°\,\mathbf{j}) = \sqrt{2}\mathbf{j}.$$

Therefore,

$$\mathbf{v} = (\mathbf{u} + \mathbf{v}) - \mathbf{u}$$
$$= \sqrt{2}\mathbf{j} - \left(\frac{\sqrt{2}}{2}\mathbf{i} + \frac{\sqrt{2}}{2}\mathbf{j}\right) = -\frac{\sqrt{2}}{2}\mathbf{i} + \frac{\sqrt{2}}{2}\mathbf{j}.$$

**69.** Let the three forces be represented by $\mathbf{F}_1$, $\mathbf{F}_2$, and $\mathbf{F}_3$ respectively.

$$\mathbf{F}_1 = 75(\cos 30°\,\mathbf{i} + \sin 30°\,\mathbf{j}) = \frac{75\sqrt{3}}{2}\mathbf{i} + \frac{75}{2}\mathbf{j}$$

$$\mathbf{F}_2 = 100(\cos 45°\,\mathbf{i} + \sin 45°\,\mathbf{j}) = 50\sqrt{2}\mathbf{i} + 50\sqrt{2}\mathbf{j}$$

$$\mathbf{F}_3 = 125(\cos 120°\,\mathbf{i} + \sin 120°\,\mathbf{j}) = -\frac{125}{2}\mathbf{i} + \frac{125\sqrt{3}}{2}\mathbf{j}$$

$$\mathbf{F}_1 + \mathbf{F}_2 + \mathbf{F}_3 = \left(\frac{75\sqrt{3}}{2} + 50\sqrt{2} - \frac{125}{2}\right)\mathbf{i} + \left(\frac{75}{2} + 50\sqrt{2} + \frac{125\sqrt{3}}{2}\right)\mathbf{j}$$

$$\|\mathbf{F}_1 + \mathbf{F}_2 + \mathbf{F}_3\| = \sqrt{\left(\frac{75\sqrt{3}}{2} + 50\sqrt{2} - \frac{125}{2}\right)^2 + \left(\frac{75}{2} + 50\sqrt{2} + \frac{125\sqrt{3}}{2}\right)^2} \approx 228.5 \text{ pounds}$$

$$\theta_{\mathbf{F}_1+\mathbf{F}_2+\mathbf{F}_3} = \arctan\left(\frac{\frac{75\sqrt{3}}{2} + 50\sqrt{2} - \frac{125}{2}}{\frac{75}{2} + 50\sqrt{2} + \frac{125\sqrt{3}}{2}}\right) \approx 71.3°$$

**73.** (a) Consider the vectors, $\mathbf{T}$, $\mathbf{u}$, and $\mathbf{v}$ where $\mathbf{T}$ is the force in the rope from the pole to the tether ball, $\mathbf{u}$ is the horizontal force pulling the ball away from the pole, and $\mathbf{v}$ represents the weight of the ball (see figure).

$$\mathbf{u} = \|\mathbf{u}\|\mathbf{i}$$

$$\mathbf{v} = -\mathbf{j}$$

$$\mathbf{T} = \|\mathbf{T}\|(\sin 30°)\mathbf{i} + \|\mathbf{T}\|(-\cos 30°)\mathbf{j}$$

Also, $\mathbf{T} = \mathbf{u} + \mathbf{v} = \|\mathbf{u}\|\mathbf{i} - \mathbf{j}$. Therefore,

$$-1 = \|\mathbf{T}\|(-\cos 30°)$$

$$\text{Tension} = \|\mathbf{T}\| = \frac{2}{\sqrt{3}} \approx 1.1547 \text{ pounds,}$$

and

$$\|\mathbf{u}\| = \|\mathbf{T}\|(\sin 30°) = \left(\frac{2}{\sqrt{3}}\right)\left(\frac{1}{2}\right) \approx 0.5774 \text{ pounds.}$$

(b) Replacing 30° with $\theta$ in part (a) yields

$$-1 = \|\mathbf{T}\|(-\cos \theta).$$

$$\text{Tension} = T = \|\mathbf{T}\| = \frac{1}{\cos \theta} = \sec \theta,$$

and

$$\|\mathbf{u}\| = \|\mathbf{T}\| \sin \theta = \sec \theta \sin \theta = \tan \theta.$$

(c)

| $\theta$ | 0° | 10° | 20° | 30° | 40° | 50° | 60° |
|---|---|---|---|---|---|---|---|
| $T$ | 1 | 1.0154 | 1.0642 | 1.1547 | 1.3054 | 1.5557 | 2 |
| $\|\mathbf{u}\|$ | 0 | 0.1763 | 0.3640 | 0.5774 | 0.8391 | 1.1918 | 1.7321 |

(d) The graph is shown in the figure.

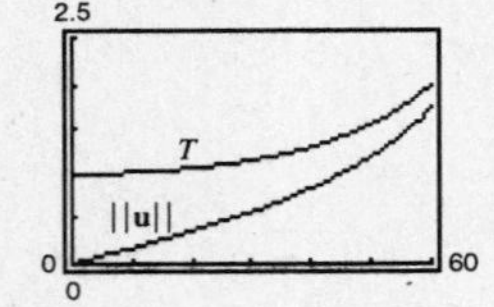


(e) Both are increasing functions for $0° \le \theta \le 60°$.

(f) $\lim_{\theta \to \pi/2} T = \lim_{\theta \to \pi/2} \sec \theta = \infty$

$\lim_{\theta \to \pi/2} \|\mathbf{u}\| = \lim_{\theta \to \pi/2} \tan \theta = \infty$

These results are expected. The forces increase without bound as the rope between the pole and the tether ball approaches the horizontal.

**77.** Let $\mathbf{u}$ represent the air speed and direction of the plane, and let $\mathbf{v}$ represent the speed and direction of the wind.

$$\mathbf{u} = 900[(-\cos 32°)\mathbf{i} + (\sin 32°)\mathbf{j}]$$

$$\mathbf{v} = 100[(\cos 45°)\mathbf{i} + (\sin 45°)\mathbf{j}]$$

$$\mathbf{u} + \mathbf{v} = (-900 \cos 32° + 100 \cos 45°)\mathbf{i} + (900 \sin 32° + 100 \sin 45°)\mathbf{j} \approx -692.53\mathbf{i} + 547.64\mathbf{j}$$

The speed of the plane is

$$\|\mathbf{u} + \mathbf{v}\| = \sqrt{(-692.53)^2 + 547.64^2} \approx 882.90 \text{ kilometers per hour.}$$

The true direction on the plane North of West is

$$\cos \theta = \frac{(\mathbf{u} + \mathbf{v}) \cdot (-\mathbf{i})}{\|\mathbf{u} + \mathbf{v}\|} \approx \frac{692.53}{882.90} \approx 0.7844$$

$$\theta \approx \arccos(0.7844) \approx 38.3° \text{ North of West.}$$

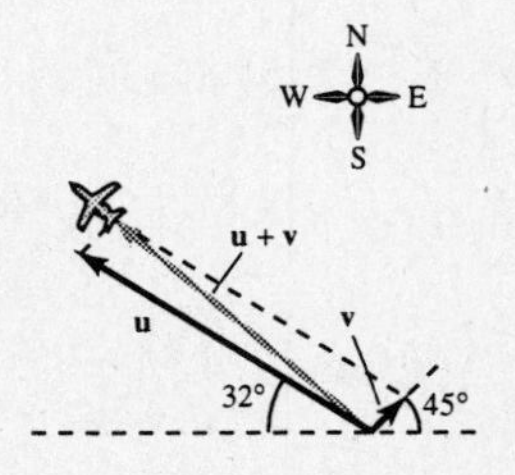

## Section 10.2 Space Coordinates and Vectors in Space

**19.** Given the three points $A(1, -3, -2)$, $B(5, -1, 2)$, and $C(-1, 1, 2)$, we have

$$|AB| = \sqrt{(5-1)^2 + (-1+3)^2 + (2+2)^2} = \sqrt{16+4+16} = \sqrt{36} = 6$$

$$|AC| = \sqrt{(-1-1)^2 + (1+3)^2 + (2+2)^2} = \sqrt{4+16+16} = \sqrt{36} = 6$$

$$|BC| = \sqrt{(-1-5)^2 + (1+1)^2 + (2-2)^2} = \sqrt{36+4} = \sqrt{40} = 2\sqrt{10}.$$

Since two sides have equal lengths, the triangle is isosceles.

**27.** The center of the circle is the midpoint of the endpoints of the diameter $(2, 0, 0)$ and $(0, 6, 0)$.

$$\text{Center: } \left(\frac{2+0}{2}, \frac{0+6}{2}, \frac{0+0}{2}\right) = (1, 3, 0)$$

The radius is the distance from the center to one of the endpoints of a diameter.

$$\text{Radius: } \sqrt{(1-2)^2 + (3-0)^2 + (0-0)^2} = \sqrt{10}$$

Since the center is $(x_0, y_0, z_0) = (1, 3, 0)$, we have

$$(x-x_0)^2 + (y-y_0)^2 + (z-z_0)^2 = r^2$$

$$(x-1)^2 + (y-3)^2 + (z-0)^2 = 10.$$

**29.** Writing the equation in standard form yields

$$x^2 + y^2 + z^2 - 2x + 6y + 8z + 1 = 0$$

$$(x^2 - 2x) + (y^2 + 6y) + (z^2 + 8z) = -1$$

$$(x^2 - 2x + 1) + (y^2 + 6y + 9) + (z^2 + 8z + 16) = -1 + 1 + 9 + 16$$

$$(x-1)^2 + (y+3)^2 + (z+4)^2 = 25.$$

Thus, the sphere is centered at $(1, -3, -4)$ with radius 5.

**39.** The initial point is $P = (0, 6, 2)$ and the terminal point is $Q = (q_1, q_2, q_3)$. Then

$$\mathbf{v} = \langle 3, -5, 6\rangle = \langle q_1 - 0, q_2 - 6, q_3 - 2\rangle.$$

Therefore,

$$q_1 - 0 = 3 \Rightarrow q_1 = 3$$

$$q_2 - 6 = -5 \Rightarrow q_2 = 1$$

$$q_3 - 2 = 6 \Rightarrow q_1 = 8,$$

and the coordinates of the terminal point are $(3, 1, 8)$.

**47.** $\mathbf{u} = \langle 1, 2, 3\rangle$, $\mathbf{w} = \langle 4, 0, -4\rangle$

$$2\mathbf{z} - 3\mathbf{u} = \mathbf{w}$$

$$2\mathbf{z} = 3\mathbf{u} + \mathbf{w}$$

$$= 3(\mathbf{i} + 2\mathbf{j} + 3\mathbf{k}) + (4\mathbf{i} - 4\mathbf{k})$$

$$= 3\mathbf{i} + 6\mathbf{j} + 9\mathbf{k} + 4\mathbf{i} - 4\mathbf{k}$$

$$= 7\mathbf{i} + 6\mathbf{j} + 5\mathbf{k}$$

$$\mathbf{z} = \tfrac{7}{2}\mathbf{i} + 3\mathbf{j} + \tfrac{5}{2}\mathbf{k} = \left\langle \tfrac{7}{2}, 3, \tfrac{5}{2}\right\rangle$$

**53.** Given the points $A(0, -2, -5)$, $B(3, 4, 4)$, and $C(2, 2, 1)$, we have

$$\overrightarrow{AB} = (3-0)\mathbf{i} + (4+2)\mathbf{j} + (4+5)\mathbf{k} = 3\mathbf{i} + 6\mathbf{j} + 9\mathbf{k}$$

$$\overrightarrow{AC} = (2-0)\mathbf{i} + (2+2)\mathbf{j} + (1+5)\mathbf{k} = 2\mathbf{i} + 4\mathbf{j} + 6\mathbf{k}$$

$$= \tfrac{2}{3}(3\mathbf{i} + 6\mathbf{j} + 9\mathbf{k}) = \tfrac{2}{3}\overrightarrow{AB}.$$

Since $\overrightarrow{AC}$ is a scaler multiple of $\overrightarrow{AB}$, the three points lie on a straight line.

**65.** $\mathbf{u} = \langle 2, -1, 2\rangle$

(a) Since the magnitude of **u** is

$$\|\mathbf{u}\| = \sqrt{2^2 + (-1)^2 + 2^2} = \sqrt{9} = 3,$$

the unit vector in the direction of **u** is

$$\frac{\mathbf{u}}{\|\mathbf{u}\|} = \frac{1}{3}\langle 2, -1, 2\rangle.$$

(b) Since the unit vector in the opposite direction is obtained by multiplying by the scalar $-1$, we have

$$(-1)\frac{\mathbf{u}}{\|\mathbf{u}\|} = -\frac{1}{3}\langle 2, -1, 2\rangle.$$

**75.** $\mathbf{u} = \langle 2, -2, 1\rangle$

$\|\mathbf{u}\| = \sqrt{2^2 + (-2)^2 + 1^2} = \sqrt{9} = 3$

Hence, the unit vector in the direction of $\mathbf{u}$ is

$$\frac{\mathbf{u}}{\|\mathbf{u}\|} = \frac{1}{3}\langle 2, -2, 1\rangle.$$

The vector $\mathbf{v}$ of length $\frac{3}{2}$ in the direction of $\mathbf{u}$ is $\frac{3}{2}$ times the unit vector in the direction of $\mathbf{u}$. Therefore,

$$\mathbf{v} = \frac{3}{2}\left[\frac{1}{3}\langle 2, -2, 1\rangle\right] = \frac{1}{2}\langle 2, -2, 1\rangle = \left\langle 1, -1, \frac{1}{2}\right\rangle.$$

**79.** Consider the points $P(4, 3, 0)$ and $Q(1, -3, 3)$. By finding the component form of the vector from $P$ to $Q$, we have

$$\overrightarrow{PQ} = \langle 1 - 4, -3 - 3, 3 - 0\rangle = \langle -3, -6, 3\rangle.$$

Let $R = (x, y, z)$ be a point on the line segment two-thirds of the way from $P$ to $Q$ (see figure). Then

$$\overrightarrow{PR} = \langle x - 4, y - 3, z - 0\rangle = \langle x - 4, y - 3, z\rangle$$

and

$$\frac{2}{3}\overrightarrow{PQ} = \frac{2}{3}\langle -3, -6, 3\rangle$$
$$= \langle -2, -4, 2\rangle = \langle x - 4, y - 3, z\rangle = \overrightarrow{PR}.$$

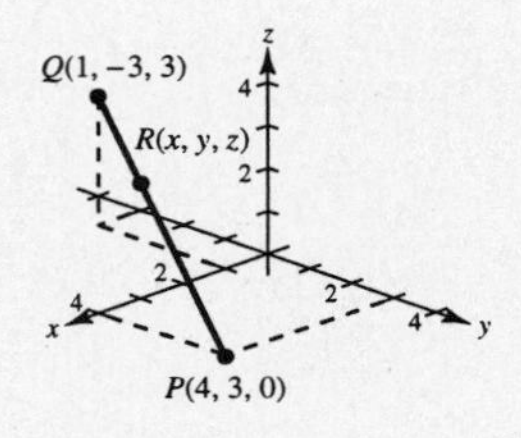


Therefore,

$$x - 4 = -2 \Rightarrow x = 2$$
$$y - 3 = -4 \Rightarrow y = -1$$
$$z = 2,$$

and the required point is $(2, -1, 2)$.

**81.** $\mathbf{u} = \mathbf{i} + \mathbf{j}, \mathbf{v} = \mathbf{j} + \mathbf{k}$

(a) See the figure.

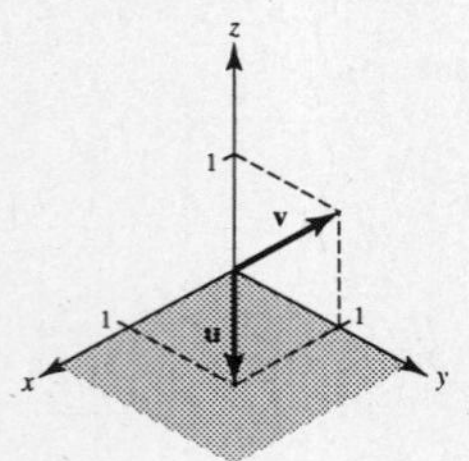


(b) $\mathbf{w} = a(\mathbf{i} + \mathbf{j}) + b(\mathbf{j} + \mathbf{k}) = \mathbf{0}$

$a\mathbf{i} + (a + b)\mathbf{j} + b\mathbf{k} = 0\mathbf{i} + 0\mathbf{j} + 0\mathbf{k}$

Therefore, $a = b = 0$.

(c) $\mathbf{w} = a(\mathbf{i} + \mathbf{j}) + b(\mathbf{j} + \mathbf{k}) = \mathbf{i} + 2\mathbf{j} + \mathbf{k}$

$a\mathbf{i} + (a + b)\mathbf{j} + b\mathbf{k} = \mathbf{i} + 2\mathbf{j} + \mathbf{k}$

Therefore, $a = b = 1$.

(d) $\mathbf{w} = a(\mathbf{i} + \mathbf{j}) + b(\mathbf{j} + \mathbf{k}) = \mathbf{i} + 2\mathbf{j} + 3\mathbf{k}$

$a\mathbf{i} + (a + b)\mathbf{j} + b\mathbf{k} = \mathbf{i} + 2\mathbf{j} + 3\mathbf{k}$

Therefore, $a = 1$, $b = 3$, and $a + b = 2$, which is a contradiction.

**83.** (a) From the Pythagorean Theorem, the distance from the center of a disk to the support in the ceiling is $\sqrt{L^2 - 18^2}$. Thus, the coordinates of the points $P$ and $Q$ in the figure are $(0, 18, 0)$ and $\left(0, 0, \sqrt{L^2 - 18^2}\right)$, respectively, and the vector from $P$ to $Q$ is $\left\langle 0, -18, \sqrt{L^2 - 18^2}\right\rangle$. The force $\mathbf{F}$ in one of the supporting wires must be a scalar multiple $c$ of the vector from $P$ to $Q$, or

$$\mathbf{F} = \left\langle 0, -18c, c\sqrt{L^2 - 18^2}\right\rangle.$$

Since there are three supporting wires, the vertical component of the force vector in one wire must be one-third the weight of the light. Therefore,

$$c\sqrt{L^2 - 18^2} = \frac{24}{3} \Rightarrow c = \frac{8}{\sqrt{L^2 - 18^2}} \quad \text{and} \quad \mathbf{F} = \left\langle 0, \frac{-8 \cdot 18}{\sqrt{L^2 - 18^2}}, 8\right\rangle.$$

The tension $T$ in each wire is

$$T = \|\mathbf{F}\| = 8\sqrt{\frac{18^2}{L^2 - 18^2} + 1} = \frac{8L}{\sqrt{L^2 - 18^2}}.$$

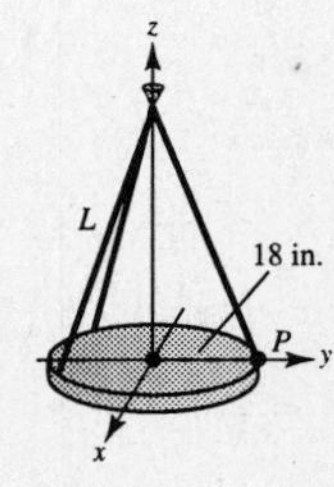


—CONTINUED—

**83. —CONTINUED—**

(b)

| $L$ | 20 | 25 | 30 | 35 | 40 | 45 | 50 |
|---|---|---|---|---|---|---|---|
| $T$ | 18.4 | 11.5 | 10.0 | 9.3 | 9.0 | 8.7 | 8.6 |

(c) The graph of the tension is shown in the figure.

Vertical asymptote: $L = 18$

Horizontal asymptote: $T = 8$

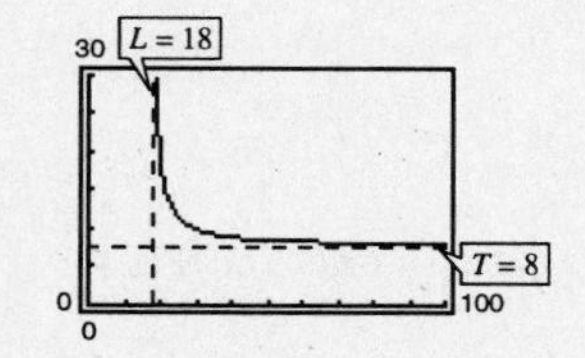


(d) $\lim_{L\to 18^+} \dfrac{8L}{\sqrt{L^2 - 18^2}} = \infty$

For $L > 0$, we can write $L = \sqrt{L^2}$. Thus, dividing the numerator and denominator by $L$ produces

$$\lim_{L\to\infty} \frac{8L}{\sqrt{L^2 - 18^2}} = \lim_{L\to\infty} \frac{\frac{8L}{L}}{\frac{\sqrt{L^2 - 18^2}}{\sqrt{L^2}}}$$

$$= \lim_{x\to 0} \frac{8}{\sqrt{1 - \frac{18^2}{L^2}}} = 8.$$

(e) Since the tension in the equation must not exceed 10 pounds, solve the following inequality.

$$\frac{8L}{\sqrt{L^2 - 18^2}} \le 10$$

$$8L \le 10\sqrt{L^2 - 18^2}$$

$$64L^2 \le 100(L^2 - 18^2)$$

$$100(18^2) \le 36L^2$$

$$\frac{100(18^2)}{36} \le L^2 \implies 30 \le L$$

Each wire must be at least 30 inches long.

## Section 10.3 The Dot Product of Two Vectors

**3.** $\mathbf{u} = \langle 2, -3, 4\rangle$, $\mathbf{v} = \langle 0, 6, 5\rangle$

(a) For the vectors $\mathbf{u} = \langle u_1, u_2, u_3\rangle$ and $\mathbf{v} = \langle v_1, v_2, v_3\rangle$,

$$\mathbf{u}\cdot\mathbf{v} = u_1v_1 + u_2v_2 + u_3v_3$$

$$= 2(0) + (-3)(6) + 4(5) = 2.$$

(b) $\mathbf{u}\cdot\mathbf{u} = 2(2) + (-3)(-3) + 4(4) = 29$

(c) $\|\mathbf{u}\|^2 = 2^2 + (-3)^2 + 4^2 = 29 = \mathbf{u}\cdot\mathbf{u}$

(d) From part (a) we have $\mathbf{u}\cdot\mathbf{v} = 2$. Therefore,

$$(\mathbf{u}\cdot\mathbf{v})\mathbf{v} = 2\mathbf{v} = \langle 0, 12, 10\rangle.$$

(e) From part (a) we have

$$\mathbf{u}\cdot(2\mathbf{v}) = 2(\mathbf{u}\cdot\mathbf{v}) = 2(2) = 4.$$

**9.** $\mathbf{u}\cdot\mathbf{v} = \|\mathbf{u}\|\,\|\mathbf{v}\|\cos\theta = (8)(5)\cos\dfrac{\pi}{3} = 40\left(\dfrac{1}{2}\right) = 20$

**17.** $\mathbf{u} = 3\mathbf{i} + 4\mathbf{j}$, $\mathbf{v} = -2\mathbf{j} + 3\mathbf{k}$

Since

$$\cos\theta = \frac{\mathbf{u}\cdot\mathbf{v}}{\|\mathbf{u}\|\,\|\mathbf{v}\|},$$

we have

$$\cos\theta = \frac{0 - 8 + 0}{\sqrt{25}\sqrt{13}} = \frac{-8\sqrt{13}}{65}$$

and

$$\theta = \arccos\left(\frac{-8\sqrt{13}}{65}\right) \approx 116.3°.$$

**23.** $\mathbf{u} = \langle 4, 3\rangle$, $\mathbf{v} = \left\langle \frac{1}{2}, -\frac{2}{3}\right\rangle$

$$\mathbf{u}\cdot\mathbf{v} = 4\left(\tfrac{1}{2}\right) + 3\left(-\tfrac{2}{3}\right) = 2 - 2 = 0$$

Therefore, $\mathbf{u}$ and $\mathbf{v}$ are orthogonal.

**31.** Assuming that $\mathbf{u}$ and $\mathbf{v}$ are nonzero vectors and $\mathbf{u} \cdot \mathbf{v} = \|\mathbf{u}\|\|\mathbf{v}\| \cos \theta$, we have the following:

(a) $\mathbf{u} \cdot \mathbf{v} = 0$ implies that $\cos \theta = 0$, or $\theta = \pi/2$.

(b) $\mathbf{u} \cdot \mathbf{v} > 0$ implies that $\cos \theta > 0$, or $0 \le \theta < \pi/2$.

(c) $\mathbf{u} \cdot \mathbf{v} < 0$ implies that $\cos \theta < 0$, or $\pi/2 < \theta \le \pi$.

**35.** $\mathbf{u} = \mathbf{i} + 2\mathbf{j} + 2\mathbf{k}$

Given a vector $\mathbf{u} = u_1\mathbf{i} + u_2\mathbf{j} + u_3\mathbf{k}$, the direction cosines are

$$\cos \alpha = \frac{u_1}{\|\mathbf{u}\|}, \ \cos \beta = \frac{u_2}{\|\mathbf{u}\|}, \ \cos \gamma = \frac{u_3}{\|\mathbf{u}\|}.$$

Therefore, for the given vector, we have

$$\cos \alpha = \frac{1}{\sqrt{1^2 + 2^2 + 2^2}} = \frac{1}{\sqrt{9}} = \frac{1}{3}$$

$$\cos \beta = \frac{2}{\sqrt{1^2 + 2^2 + 2^2}} = \frac{2}{\sqrt{9}} = \frac{2}{3}$$

$$\cos \gamma = \frac{2}{\sqrt{1^2 + 2^2 + 2^2}} = \frac{2}{\sqrt{9}} = \frac{2}{3}.$$

The sum of the squares of the direction cosines is

$$\cos^2 \alpha + \cos^2 \beta + \cos^2 \gamma = \left(\frac{1}{3}\right)^2 + \left(\frac{2}{3}\right)^2 + \left(\frac{2}{3}\right)^2 = 1.$$

**47.** (a) The projection of $\mathbf{u} = \langle 2, 1, 2 \rangle$ onto $\mathbf{v} = \langle 0, 3, 4 \rangle$ is

$$\mathbf{w}_1 = \left(\frac{\mathbf{u} \cdot \mathbf{v}}{\|\mathbf{v}\|^2}\right)\mathbf{v}$$

$$= \frac{11}{25}\langle 0, 3, 4 \rangle = \left\langle 0, \frac{33}{25}, \frac{44}{25} \right\rangle.$$

(b) The vector component of $\mathbf{u}$ orthogonal to $\mathbf{v}$ is

$$\mathbf{w}_2 = \mathbf{u} = \mathbf{w}_1 = \left\langle 2, -\frac{8}{25}, \frac{6}{25} \right\rangle.$$

**57.** Let $\mathbf{v} = \langle v_1, v_2, v_3 \rangle$ be a vector orthogonal to $\mathbf{u} = \langle 3, 1, -2 \rangle$. Then

$$\mathbf{u} \cdot \mathbf{v} = 3v_1 + v_2 - 2v_3 = 0.$$

Solving the equation for $v_2$ produces

$$v_2 = 2v_3 - 3v_1.$$

If we let $v_1 = 1$ and $v_3 = 2$, then $v_2 = 1$. Therefore, a vector orthogonal to $\mathbf{u}$ is $\mathbf{v} = \langle 1, 1, 2 \rangle$. A vector in the opposite direction is $-\mathbf{v} = \langle -1, -1, -2 \rangle$. (Note that the answers are not unique since our choices for $v_1$ and $v_3$ in the equation above were arbitrary.)

**61.** The work done by a constant force $\mathbf{F}$ as its point of application moves along the vector $\overrightarrow{PQ}$ is given by

$$W = \mathbf{F} \cdot \overrightarrow{PQ}.$$

From the figure it follows that

$$\mathbf{F} = 85[(\cos 60°)\mathbf{i} + (\sin 60°)\mathbf{j}]$$

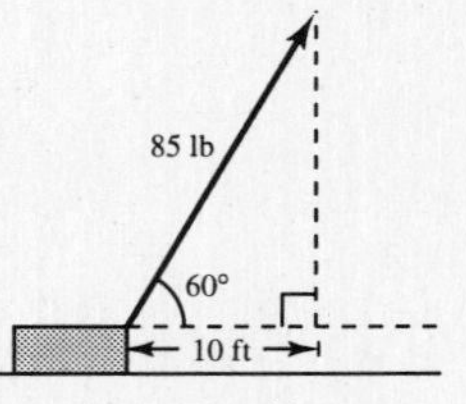


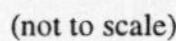
(not to scale)

and

$$\overrightarrow{PQ} = 10\mathbf{i}.$$

Therefore,

$$W = 85[(\cos 60°)(10) + (\sin 60°)(0)] = 425 \text{ ft} \cdot \text{lb}.$$

## Section 10.4 The Cross Product of Two Vectors in Space

**11.** $\mathbf{u} = \langle 1, 1, 1 \rangle, \mathbf{v} = \langle 2, 1, -1 \rangle$

The cross product is given by

$$\mathbf{u} \times \mathbf{v} = \begin{vmatrix} \mathbf{i} & \mathbf{j} & \mathbf{k} \\ 1 & 1 & 1 \\ 2 & 1 & -1 \end{vmatrix} = \mathbf{i}\begin{vmatrix} 1 & 1 \\ 1 & -1 \end{vmatrix} - \mathbf{j}\begin{vmatrix} 1 & 1 \\ 2 & -1 \end{vmatrix} + \mathbf{k}\begin{vmatrix} 1 & 1 \\ 2 & 1 \end{vmatrix}$$

$$= (-1 - 1)\mathbf{i} - (-1 - 2)\mathbf{j} + (1 - 2)\mathbf{k}$$

$$= -2\mathbf{i} + 3\mathbf{j} - \mathbf{k} = \langle -2, 3, -1 \rangle.$$

Using the dot product yields

$$\mathbf{u} \cdot (\mathbf{u} \times \mathbf{v}) = (\mathbf{i} + \mathbf{j} + \mathbf{k}) \cdot (-2\mathbf{i} + 3\mathbf{j} - \mathbf{k}) = -2 + 3 - 1 = 0$$

and

$$\mathbf{v} \cdot (\mathbf{u} \times \mathbf{v}) = (2\mathbf{i} + \mathbf{j} - \mathbf{k}) \cdot (-2\mathbf{i} + 3\mathbf{j} - \mathbf{k}) = -4 + 3 + 1 = 0.$$

Therefore $\mathbf{u} \times \mathbf{v}$ is orthogonal to both $\mathbf{u}$ and $\mathbf{v}$.

**23.** The area of the parallelogram with adjacent sides $\mathbf{u} = \mathbf{j}$ and $\mathbf{v} = \mathbf{j} + \mathbf{k}$ is the magnitude of their cross product.

$$\mathbf{u} \times \mathbf{v} = \begin{vmatrix} \mathbf{i} & \mathbf{j} & \mathbf{k} \\ 0 & 1 & 0 \\ 0 & 1 & 1 \end{vmatrix} = \mathbf{i}$$

$$A = \|\mathbf{u} \times \mathbf{v}\| = \|\mathbf{i}\| = 1$$

**27.** If you denote the four given points by $A(1, 1, 1)$, $B(2, 3, 4)$, $C(6, 5, 2)$, and $D(7, 7, 5)$, respectively, the vectors representing the sides of the quadrilateral are

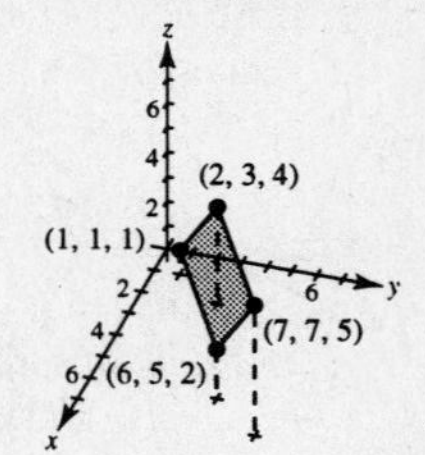


$$\overrightarrow{AB} = (2 - 1)\mathbf{i} + (3 - 1)\mathbf{j} + (4 - 1)\mathbf{k} = \mathbf{i} + 2\mathbf{j} + 3\mathbf{k}$$

$$\overrightarrow{AC} = (6 - 1)\mathbf{i} + (5 - 1)\mathbf{j} + (2 - 1)\mathbf{k} = 5\mathbf{i} + 4\mathbf{j} + \mathbf{k}$$

$$\overrightarrow{CD} = (7 - 6)\mathbf{i} + (7 - 5)\mathbf{j} + (5 - 2)\mathbf{k} = \mathbf{i} + 2\mathbf{j} + 3\mathbf{k}$$

$$\overrightarrow{DB} = (7 - 2)\mathbf{i} + (7 - 3)\mathbf{j} + (5 - 4)\mathbf{k} = 5\mathbf{i} + 4\mathbf{j} + \mathbf{k}$$

Since $\overrightarrow{AB} = \overrightarrow{CD}$ and $\overrightarrow{AC} = \overrightarrow{DB}$, the quadrilateral is a parallelogram. The area of the parallelogram is given by $\|\overrightarrow{AB} \times \overrightarrow{AC}\|$.

$$\overrightarrow{AB} \times \overrightarrow{AC} = \begin{vmatrix} \mathbf{i} & \mathbf{j} & \mathbf{k} \\ 1 & 2 & 3 \\ 5 & 4 & 1 \end{vmatrix} = \begin{vmatrix} 2 & 3 \\ 4 & 1 \end{vmatrix}\mathbf{i} - \begin{vmatrix} 1 & 3 \\ 5 & 1 \end{vmatrix}\mathbf{j} + \begin{vmatrix} 1 & 2 \\ 5 & 4 \end{vmatrix}\mathbf{k}$$

$$= -10\mathbf{i} + 14\mathbf{j} - 6\mathbf{k} = 2(-5\mathbf{i} + 7\mathbf{j} - 3\mathbf{k})$$

$$\|\overrightarrow{AB} \times \overrightarrow{AC}\| = 2\sqrt{25 + 49 + 9} = 2\sqrt{83}.$$

**35.** $\mathbf{u} = \langle 2, 0, 1 \rangle, \mathbf{v} = \langle 0, 3, 0 \rangle, \mathbf{w} = \langle 0, 0, 1 \rangle$

$$\mathbf{v} \times \mathbf{w} = \begin{vmatrix} \mathbf{i} & \mathbf{j} & \mathbf{k} \\ 0 & 3 & 0 \\ 0 & 0 & 1 \end{vmatrix} = \mathbf{i}\begin{vmatrix} 3 & 0 \\ 0 & 1 \end{vmatrix} - \mathbf{j}\begin{vmatrix} 0 & 0 \\ 0 & 1 \end{vmatrix} + \mathbf{k}\begin{vmatrix} 0 & 3 \\ 0 & 0 \end{vmatrix} = 3\mathbf{i}$$

Therefore, the triple scalar product is

$$\mathbf{u} \cdot (\mathbf{v} \times \mathbf{w}) = 2(3) + 0(0) + 1(0) = 6.$$

**39.** The vertices of a parallelepiped are (0, 0, 0), (3, 0, 0), (0, 5, 1) (3, 5, 1), (2, 0, 5), (5, 0, 5), (2, 5, 6), and (5, 5, 6). From the figure, it can be seen that three adjacent sides of the parallelepiped are given by **u**, **v**, and **w** where

$\mathbf{u} = 3\mathbf{i}$, $\mathbf{v} = 5\mathbf{j} + \mathbf{k}$, and $\mathbf{w} = 2\mathbf{i} + 5\mathbf{k}$.

Since the volume of the parallelepiped is the absolute value of the triple scalar product $\mathbf{u} \cdot (\mathbf{v} \times \mathbf{w})$, we have

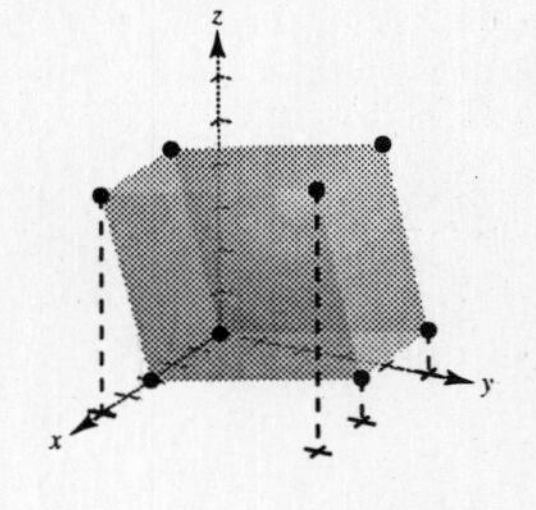


$$\mathbf{v} \times \mathbf{w} = \begin{vmatrix} \mathbf{i} & \mathbf{j} & \mathbf{k} \\ 0 & 5 & 1 \\ 2 & 0 & 5 \end{vmatrix} = \begin{vmatrix} 5 & 1 \\ 0 & 5 \end{vmatrix}\mathbf{i} - \begin{vmatrix} 0 & 1 \\ 2 & 5 \end{vmatrix}\mathbf{j} + \begin{vmatrix} 0 & 5 \\ 2 & 0 \end{vmatrix}\mathbf{k}$$

$$= 25\mathbf{i} + 2\mathbf{j} - 10\mathbf{k}$$

$$|\mathbf{u} \cdot (\mathbf{v} \times \mathbf{w})| = |3(25) + 0(2) + 0(-10)| = 75.$$

**41.** If we represent the 20 pound force as $\mathbf{F} = -20\mathbf{k}$ and the lever of length one-half foot as

$$\mathbf{v} = \tfrac{1}{2}(\cos 40^\circ\, \mathbf{j} + \sin 40^\circ\, \mathbf{k}),$$

then the moment of **F** about $P$ is given by

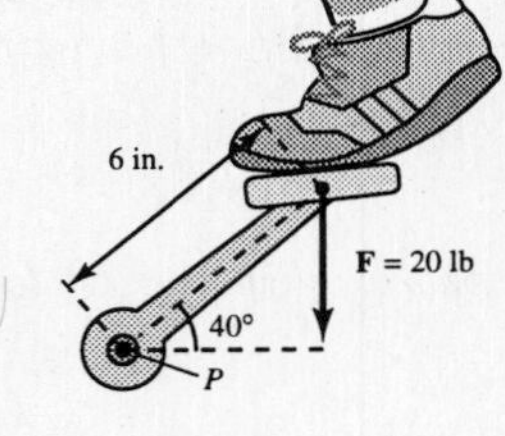


$$\mathbf{M} = \mathbf{v} \times \mathbf{F} = \begin{vmatrix} \mathbf{i} & \mathbf{j} & \mathbf{k} \\ 0 & \frac{1}{2}\cos 40^\circ & \frac{1}{2}\sin 40^\circ \\ 0 & 0 & -20 \end{vmatrix} = -10\cos 40^\circ\, \mathbf{i}.$$

The torque is the magnitude of this moment. Thus,

$$\text{torque} = \|\mathbf{M}\| = 10\cos 40^\circ \approx 7.66 \text{ foot-pounds}.$$

## Section 10.5 Lines and Planes in Space

**9.** (a) If $A\left(-\frac{2}{3}, \frac{2}{3}, 1\right)$ and $B(5, -3, -2)$, then a direction vector for the line passing through these points is given by

$$\overrightarrow{AB} = \frac{17}{3}\mathbf{i} - \frac{11}{3}\mathbf{j} - 3\mathbf{k} = \frac{1}{3}(17\mathbf{i} - 11\mathbf{j} - 9\mathbf{k})$$

and a set of direction numbers for the line is $a = 17$, $b = -11$, and $c = -9$. Using the form

$$x = x_1 + at,\ y = y_1 + bt,\ z = z_1 + ct$$

with $(x_1, y_1, z_1) = (5, -3, -2)$, a set of parametric equation for the line is

$$x = 5 + 17t,\ y = -3 - 11t,\ z = -2 - 9t.$$

(b) Solving for $t$ in each equation of part (a) yields

$$t = \frac{x-5}{17} = \frac{y+3}{-11} = \frac{z+2}{-9}.$$

Consequently, the symmetric form is

$$\frac{x-5}{17} = \frac{y+3}{-11} = \frac{z+2}{-9}.$$

**15.** $x = 4t + 2$ $\qquad$ $x = 2s + 2$

$y = 3$ $\qquad$ $y = 2s + 3$

$z = -t + 1$ $\qquad$ $z = s + 1$

At the point of intersection, the coordinates for one line equal the corresponding coordinates for the other line. Thus, we have three equations.

(1) $\quad 4t + 2 = 2s + 2$

(2) $\quad 3 = 2s + 3$

(3) $\quad -t + 1 = s + 1$

From equation (2) it follows that $s = 0$ and consequently, from equation (3), $t = 0$. Letting $s = t = 0$, equation (1) is satisfied and we can conclude that the two lines intersect. Substituting zero for $s$ or for $t$, we obtain the point (2, 3, 1).

To find the cosine of the angle of intersection, consider the vectors

$$\mathbf{u} = 4\mathbf{i} - \mathbf{k} \quad \text{and} \quad \mathbf{v} = 2\mathbf{i} + 2\mathbf{j} + \mathbf{k}$$

that have the respective directions of the two given lines. Therefore,

$$\cos\theta = \frac{|\mathbf{u} \cdot \mathbf{v}|}{\|\mathbf{u}\|\,\|\mathbf{v}\|} = \frac{8-1}{\sqrt{17}\sqrt{9}} = \frac{7}{3\sqrt{17}} = \frac{7\sqrt{17}}{51}.$$

**25.** The plane containing the point $(x_1, y_1, z_1)$ and having a normal vector $\mathbf{n} = a\mathbf{i} + b\mathbf{j} + c\mathbf{k}$ can be represented by the equation

$$a(x - x_1) + b(y - y_1) + c(z - z_1) = 0.$$

Point: $(x_1, y_1, z_1) = (3, 2, 2)$

Normal vector: $\mathbf{n} = a\mathbf{i} + b\mathbf{j} + c\mathbf{k} = 2\mathbf{i} + 3\mathbf{j} - \mathbf{k}$

Therefore, the equation of the plane is

$$2(x - 3) + 3(y - 2) - 1(z - 2) = 0$$
$$2x + 3y - z = 10.$$

**29.** $(0, 0, 0), (1, 2, 3), (-2, 3, 3)$

To use the form

$$a(x - x_1) + b(y - y_1) + c(z - z_1) = 0$$

we need to know a point in the plane and a vector $\mathbf{n}$ that is normal to the plane. To obtain a normal vector, use the cross product of the vectors $\mathbf{v}_1$ and $\mathbf{v}_2$ from the point $(0, 0, 0)$ to $(1, 2, 3)$ and to $(-2, 3, 3)$, respectively. We have

$$\mathbf{v}_1 = \mathbf{i} + 2\mathbf{j} + 3\mathbf{k} \quad \text{and} \quad \mathbf{v}_2 = -2\mathbf{i} + 3\mathbf{j} + 3\mathbf{k}.$$

Thus, the vector

$$\mathbf{n} = \mathbf{v}_1 \times \mathbf{v}_2 = \begin{vmatrix} \mathbf{i} & \mathbf{j} & \mathbf{k} \\ 1 & 2 & 3 \\ -2 & 3 & 3 \end{vmatrix} = -3\mathbf{i} - 9\mathbf{j} + 7\mathbf{k}$$

is normal to the given plane. Using the direction numbers from $\mathbf{n}$ and the point $(0, 0, 0)$ in the plane, we have

$$-3(x - 0) - 9(y - 0) + 7(z - 0) = 0$$
$$3x + 9y - 7z = 0.$$

**35.** $\dfrac{x - 1}{-2} = y - 4 = z$ and $\dfrac{x - 2}{-3} = \dfrac{y - 1}{4} = \dfrac{z - 2}{-1}$.

Writing the equation of the lines in parametric form, we have

$$x = 1 - 2t \qquad x = 2 - 3s$$
$$y = 4 + t \qquad y = 1 + 4s$$
$$z = t \qquad z = 2 - s$$

To find the point of intersection of the lines, we equate the expressions for the respective values of $x$, $y$, and $z$, and solve the resulting equations. The solution is $t = s = 1$, and the point of intersection is $(-1, 5, 1)$. Since the direction vectors of the lines are

$$\mathbf{v}_1 = -2\mathbf{i} + \mathbf{j} + \mathbf{k} \quad \text{and} \quad \mathbf{v}_2 = -3\mathbf{i} + 4\mathbf{j} - \mathbf{k},$$

the vector $\mathbf{n}$ normal to the plane is

$$\mathbf{n} = \mathbf{v}_1 \times \mathbf{v}_2 = \begin{vmatrix} \mathbf{i} & \mathbf{j} & \mathbf{k} \\ -2 & 1 & 1 \\ -3 & 4 & -1 \end{vmatrix} = -5(\mathbf{i} + \mathbf{j} + \mathbf{k}).$$

Therefore, the equation of the plane is

$$1(x + 1) + 1(y - 5) + 1(z - 1) = 0$$
$$x + y + z = 5.$$

**37.** Let $\mathbf{v}$ be the vector from $(-1, 1, -1)$ to $(2, 2, 1)$, let $\mathbf{n}$ be a vector normal to the plane $2x - 3y + z = 3$. Then $\mathbf{v}$ and $\mathbf{n}$ both lie in the required plane, where

$$\mathbf{v} = 3\mathbf{i} + \mathbf{j} + 2\mathbf{k} \quad \text{and} \quad \mathbf{n} = 2\mathbf{i} - 3\mathbf{j} + \mathbf{k}.$$

The vector

$$\mathbf{v} \times \mathbf{n} = \begin{vmatrix} \mathbf{i} & \mathbf{j} & \mathbf{k} \\ 3 & 1 & 2 \\ 2 & -3 & 1 \end{vmatrix} = 7\mathbf{i} + \mathbf{j} - 11\mathbf{k}$$

is normal to the required plane. Finally, since the point $(2, 2, 1)$ lies in the plane, an equation for the plane is

$$7(x - 2) + 1(y - 2) - 11(z - 1) = 0$$
$$7x + y - 11z = 5.$$

**43.** Vectors normal to the planes $x - 3y + 6z = 4$ and $5x + y - z = 4$ are $\mathbf{n}_1 = \langle 1, -3, 6 \rangle$ and $\mathbf{n}_2 = \langle 5, 1, -1 \rangle$, respectively. Since there is no scalar $c$ such that $\mathbf{n}_1 = c\mathbf{n}_2$, the vectors, and thus the planes, are not parallel. Also, since $\mathbf{n}_1 \cdot \mathbf{n}_2 \neq 0$, the vectors, and thus the planes, are not orthogonal. The cosine of the angle $\theta$ between the two planes is given by

$$\cos\theta = \frac{|\mathbf{n}_1 \cdot \mathbf{n}_2|}{\|\mathbf{n}_1\|\|\mathbf{n}_2\|} = \frac{|5 - 3 - 6|}{\sqrt{46}\sqrt{27}} = \frac{4\sqrt{138}}{414}.$$

Therefore,

$$\theta = \arccos\left(\frac{4\sqrt{138}}{414}\right) \approx 83.5°.$$

**57.** $3x + 2y - z = 7,\ x - 4y + 2z = 0$

Let $\mathbf{n}_1 = 3\mathbf{i} + 2\mathbf{j} - \mathbf{k}$ and $\mathbf{n}_2 = \mathbf{i} - 4\mathbf{j} + 2\mathbf{k}$ be the normal vectors to the respective planes. The line of intersection of the two planes will have the same direction as the vector $\mathbf{n}_1 \times \mathbf{n}_2$. Since

$$\mathbf{n}_1 \times \mathbf{n}_2 = \begin{vmatrix} \mathbf{i} & \mathbf{j} & \mathbf{k} \\ 3 & 2 & -1 \\ 1 & -4 & 2 \end{vmatrix}$$

$$= 0\mathbf{i} - 7\mathbf{j} - 14\mathbf{k} = -7(0\mathbf{i} + \mathbf{j} + 2\mathbf{k}),$$

a set of direction numbers for the line of intersections is 0, 1, and 2. By solving the equations for the two planes simultaneously, we can find points on the line of intersection.

$$3x + 2y - z = 7 \Rightarrow 6x + 4y - 2z = 14$$

$$x - 4y + 2z = 0 \Rightarrow \ \ x - 4y + 2z = 0$$

$$7x = 14 \text{ or } x = 2$$

By substituting 2 for $x$, we obtain the equation $2y - z = 1$. If we let $y = 1$, then $z = 1$. Thus, $(2, 1, 1)$ lies on the line of intersection, and we conclude that a set of parametric equations for the line of intersection is

$$x = 2,\ y = 1 + t,\ z = 1 + 2t.$$

**59.** Line: $\dfrac{x - (1/2)}{1} = \dfrac{y + (3/2)}{-1} = \dfrac{z + 1}{2}$

Plane: $2x - 2y + z = 12$

The parametric equations for the line are

$$x = \frac{1}{2} + t,\ y = \frac{-3}{2} - t,\ z = -1 + 2t.$$

If the line intersects the plane, then the expressions for $x$, $y$, and $z$ from the line must satisfy the equation of the plane. Thus,

$$2x - 2y + z = 12$$

$$2\left(\frac{1}{2} + t\right) - 2\left(\frac{-3}{2} - t\right) + (-1 + 2t) = 12$$

$$6t + 3 = 12$$

$$6t = 9 \Rightarrow t = \frac{3}{2}$$

and we conclude that the point of intersection occurs when $t = \frac{3}{2}$. This yields the point $(2, -3, 2)$.

**67.** The distance between a point $Q$ and a line in space is

$$D = \frac{\|\overrightarrow{PQ} \times \mathbf{u}\|}{\|\mathbf{u}\|}$$

where $\mathbf{u}$ is the direction vector for the line and $P$ is a point on the line. For the line $x = 4t - 2$, $y = 3$, and $z = -t + 1$, the direction vector is $\mathbf{u} = \langle 4, 0, -1\rangle$ and a point $P$ on the line is $(-2, 3, 1)$ when $t = 0$. Since $Q = (1, 5, -2)$,

$$\overrightarrow{PQ} = \langle 3, 2, -3\rangle.$$

$$\overrightarrow{PQ} \times \mathbf{u} = \begin{vmatrix} \mathbf{i} & \mathbf{j} & \mathbf{k} \\ 3 & 2 & -3 \\ 4 & 0 & -1 \end{vmatrix} = \langle -2, -9, -8\rangle$$

$$D = \frac{\|\overrightarrow{PQ} \times \mathbf{u}\|}{\|\mathbf{u}\|} = \frac{\sqrt{(-2)^2 + (-9)^2 + (-8)^2}}{\sqrt{4^2 + 0^2 + (-1)^2}} = \frac{\sqrt{2533}}{17} \approx 2.96$$

**73.** Vectors for two adjacent edges of one side of the chute shown in the figure are $\langle 6, 0, 0\rangle$ and $\langle -1, -1, 8\rangle$, and a normal vector to the side is

$$\mathbf{n}_1 = \begin{vmatrix} \mathbf{i} & \mathbf{j} & \mathbf{k} \\ 6 & 0 & 0 \\ -1 & -1 & 8 \end{vmatrix} = -48\mathbf{j} - 6\mathbf{k}.$$

Vectors for two adjacent edges of the adjacent side of the chute shown in the figure are $\langle 0, 6, 0\rangle$ and $\langle -1, -1, 8\rangle$, and a normal vector to the side is

$$\mathbf{n}_2 = \begin{vmatrix} \mathbf{i} & \mathbf{j} & \mathbf{k} \\ 0 & 6 & 0 \\ -1 & -1 & 8 \end{vmatrix} = 48\mathbf{i} + 6\mathbf{k}.$$

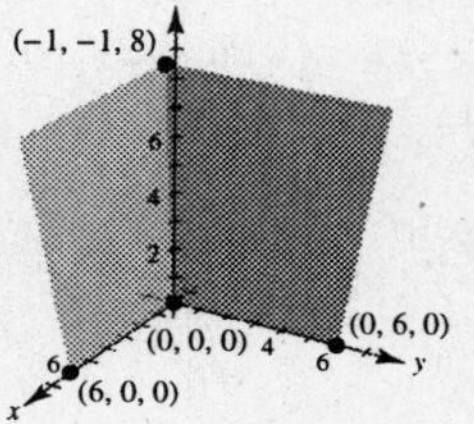


The angle $\theta$ between the two adjacent sides is the angle between their normal vectors.

$$\cos\theta = \frac{|\mathbf{n}_1 \cdot \mathbf{n}_2|}{\|\mathbf{n}_1\|\|\mathbf{n}_2\|} = \frac{36}{2340} = \frac{1}{65}$$

$$\theta = \arccos\frac{1}{65} \approx 89.1^\circ$$

## Section 10.6 Surfaces in Space

**11.** $x_2 - y = 0$

Since the $z$-coordinate is missing in the equation, the surface is a cylindrical surface with rulings parallel to the $z$-axis. The generating curve is the parabola $y = x^2$ and the surface is called a *parabolic cylinder.* (see figure.)

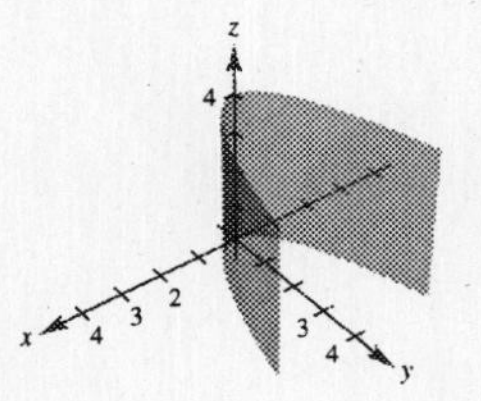


**21.** $16x^2 - y^2 + 16z^2 = 4$

The equation has the form

$$\frac{x^2}{1/4} - \frac{y^2}{4} + \frac{z^2}{1/4} = 1$$

which is the form for a **hyperboloid of one sheet.** The axis of the hyperboloid is the $y$-axis. The $xz$-trace $(y = 0)$ is the circle

$$\frac{x^2}{1/4} + \frac{z^2}{1/4} = 1$$

and the $xy$- and $yz$- traces are the hyperbolas

$$\frac{x^2}{1/4} - \frac{y^2}{4} = 1 \quad \text{and} \quad \frac{z^2}{1/4} - \frac{y^2}{4} = 1. \qquad \text{(see figure)}$$

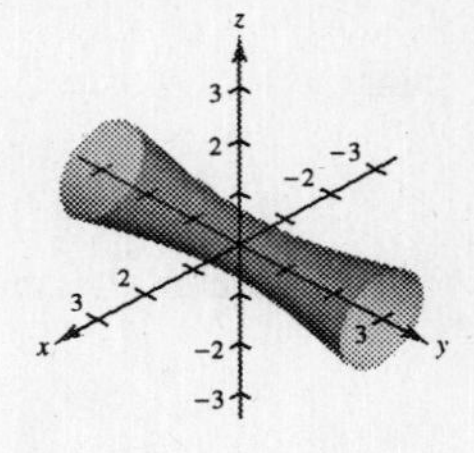


**25.** $x^2 - y^2 + z = 0$

The given equation can be written as

$$z = -\frac{x^2}{1} + \frac{y^2}{1}$$

which is the form for a **hyperbolic paraboloid.**

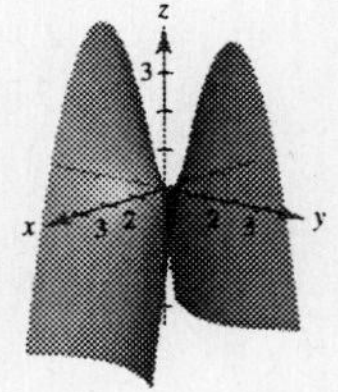


$xy$-trace $(z = 0)$: $y = \pm x$ (intersecting lines)

$xz$-trace $(y = 0)$: $z = -x^2$ (parabola opening downward)

$yz$-trace $(x = 0)$: $z = 13y^2$ (parabola opening upward)

Traces parallel to the $xy$-coordinate plane are hyperbolas. For example, when $z = 1$, we have $y^2 - x^2 = 1$.

**41.** The surface $z = 2\sqrt{x^2 + y^2}$ is a cone and the graph of $z = 2$ is a plane 2 units above the $xy$-plane. The intersection of the surfaces is a circle of radius 1.

$$2\sqrt{x^2 + y^2} = 2$$

$$\sqrt{x^2 + y^2} = 1$$

$$x^2 + y^2 = 1$$

**47.** Since we are revolving the curve $z = 2y$ in the $yz$-plane about the $z$-axis, the equation for the surface of revolution has the form $x^2 + y^2 = [a(z)]^2$ where $y = a(z) = z/2$. Therefore, the equation is

$$x^2 + y^2 = \frac{z^2}{4} \quad \text{or} \quad 4x^2 + 4y^2 = z^2.$$

**53.** The distance from the center of the rectangle to the axis of revolution is $p(x) = x$, and the height of the rectangle is $h(x) = 4x - x^2$.

$$\begin{aligned} V &= 2\pi\int_a^b p(x)h(x)\,dx \\ &= 2\pi\int_0^4 x(4x - x^2)\,dx \\ &= 2\pi\int_0^4 (4x^2 - x^3)\,dx = 2\pi\left[\frac{4}{3}x^3 - \frac{1}{4}x^4\right]_0^4 = \frac{128\pi}{3} \end{aligned}$$

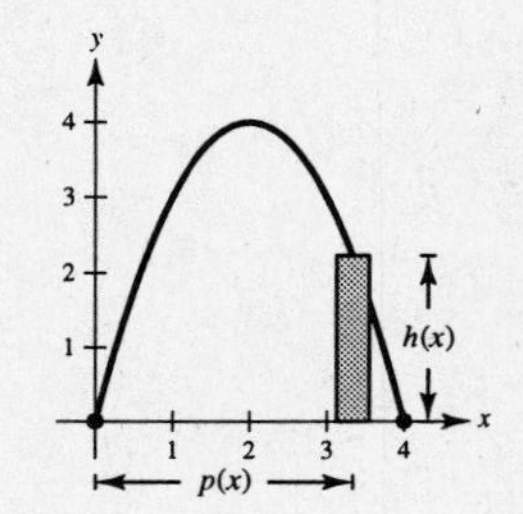


## Section 10.7 Cylindrical and Spherical Coordinates

**5.** Rectangular coordinates: $(2, -2, -4)$

Since $x = 2$, $y = -2$, and $z = -4$, we have

$$r = \pm\sqrt{x^2 + y^2} = \pm\sqrt{4 + 4} = \pm 2\sqrt{2}$$

$$\theta = \arctan\frac{y}{x} = \arctan\frac{-2}{2} = \arctan(-1) = -\frac{\pi}{4} \text{ or } \frac{3\pi}{4}$$

$$z = -4.$$

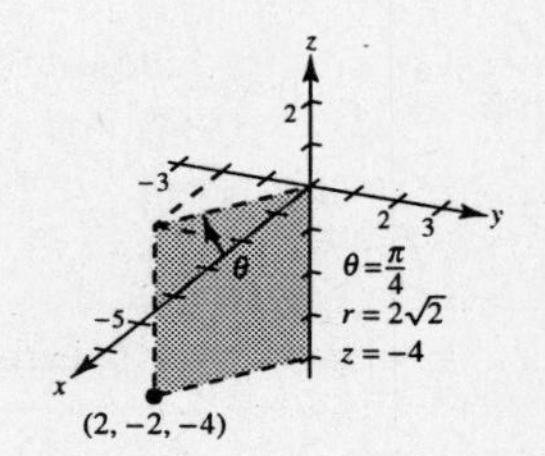


Therefore, one set of corresponding cylindrical coordinates for the given point is $(r, \theta, z) = (2\sqrt{2}, -\pi/4, -4)$. (see figure.) Another set of cylindrical coordinates for the given point is $(-2\sqrt{2}, 3\pi/4, -4)$.

**11.** Cylindrical coordinates: $\left(4, \frac{7\pi}{6}, 3\right)$

Since $\left(4, \frac{7\pi}{6}, 3\right) = (r, \theta, z)$ we have

$$x = r\cos\theta = 4\cos\frac{7\pi}{6} = -2\sqrt{3}$$

$$y = r\sin\theta = 4\sin\frac{7\pi}{6} = -2$$

$$z = 3.$$

Therefore, in rectangular coordinates, the point is $(-2\sqrt{3}, -2, 3)$.

**17.** Cylindrical coordinates: $r = 2\sin\theta$

Since $x^2 + y^2 = r^2$ and $y = r\sin\theta$, we have

$$\begin{aligned} r &= 2\sin\theta \\ r^2 &= 2r\sin\theta \\ x^2 + y^2 &= 2y \\ x^2 + y^2 - 2y + 1 &= 1 \\ x^2 + (y - 1)^2 &= 1. \end{aligned}$$

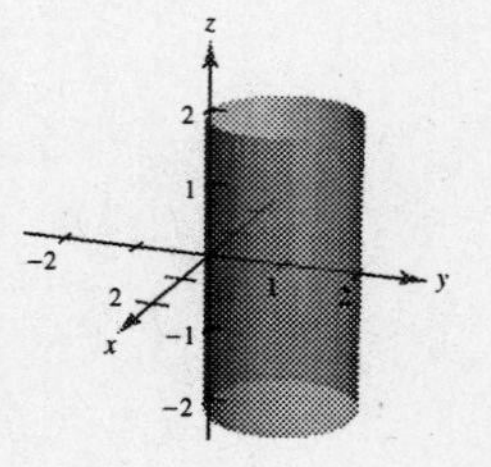


Therefore, the graph of the equation is a circular cylinder with rulings parallel to the $z$-axis.

**23.** Rectangular coordinates: $(-2, 2\sqrt{3}, 4)$

Since $(-2, 2\sqrt{3}, 4) = (x, y, z)$, we have

$$\rho = \sqrt{x^2 + y^2 + z^2} = \sqrt{4 + 12 + 16} = 4\sqrt{2}$$

$$\theta = \arctan \frac{y}{x} = \arctan \frac{2\sqrt{3}}{-2} = \frac{2\pi}{3}$$

$$\phi = \arccos \frac{z}{\rho} = \arccos \frac{4}{4\sqrt{2}} = \frac{\pi}{4}.$$

Therefore, in spherical coordinates, the point is $\left(4\sqrt{2}, \frac{2\pi}{3}, \frac{\pi}{4}\right)$.

**27.** Spherical coordinates: $\left(4, \frac{\pi}{6}, \frac{\pi}{4}\right)$

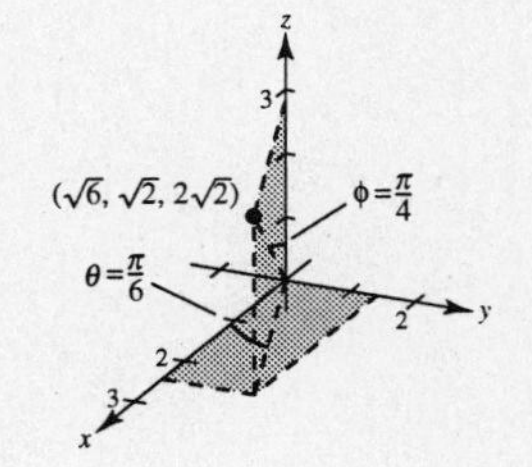


Since $\left(4, \frac{\pi}{6}, \frac{\pi}{4}\right) = (\rho, \theta, \phi)$, we have

$$x = \rho \sin \phi \cos \theta = 4 \sin \frac{\pi}{4} \cos \frac{\pi}{6} = 4\left(\frac{\sqrt{2}}{2}\right)\left(\frac{\sqrt{3}}{2}\right) = \sqrt{6}$$

$$y = \rho \sin \phi \sin \theta = 4 \sin \frac{\pi}{4} \sin \frac{\pi}{6} = 4\left(\frac{\sqrt{2}}{2}\right)\left(\frac{1}{2}\right) = \sqrt{2}$$

$$z = \rho \cos \phi = 4 \cos \frac{\pi}{4} = 4\left(\frac{\sqrt{2}}{2}\right) = 2\sqrt{2}.$$

Therefore, in rectangular coordinates, the point is $(\sqrt{6}, \sqrt{2}, 2\sqrt{2})$.

**39.** Spherical coordinates: $\rho = 4 \cos \phi$

Since

$$\cos \phi = \frac{z}{\sqrt{x^2 + y^2 + z^2}},$$

write

$$\frac{\rho}{4} = \cos \phi = \frac{z}{\sqrt{x^2 + y^2 + z^2}}.$$

Furthermore, since $\rho = \sqrt{x^2 + y^2 + z^2}$, it follows that

$$\frac{\sqrt{x^2 + y^2 + z^2}}{4} = \frac{z}{\sqrt{x^2 + y^2 + z^2}}$$

$$x^2 + y^2 + z^2 = 4z$$

$$x^2 + y^2 + z^2 - 4z = 0.$$

Completing the square on variable $z$ yields the equation

$$x^2 + y^2 + (z - 2)^2 = 4$$

which represents a sphere of radius 2, centered at $(0, 0, 2)$.

**45.** Cylindrical coordinates: $\left(4, -\frac{\pi}{6}, 6\right)$

Since $\left(4, -\frac{\pi}{6}, 6\right) = (r, \theta, z)$, we have

$$\rho^2 = x^2 + y^2 + z^2 = r^2 + z^2 = 16 + 36 = 52$$

or

$$\rho = 2\sqrt{13} > 0.$$

Furthermore,

$$\cos \phi = \frac{z}{\sqrt{x^2 + y^2 + z^2}}$$

$$= \frac{z}{\sqrt{r^2 + z^2}} = \frac{6}{\sqrt{52}} = \frac{3}{\sqrt{13}}$$

or

$$\phi = \arccos \frac{3}{\sqrt{13}}.$$

Therefore, in spherical coordinates, the point is

$$\left(2\sqrt{13}, -\frac{\pi}{6}, \arccos \frac{3}{\sqrt{13}}\right).$$

**53.** Spherical coordinates: $(\rho, \theta, \phi) = \left(8, \frac{7\pi}{6}, \frac{\pi}{6}\right)$

$$r = \rho \sin \phi = 8 \sin \frac{\pi}{6} = 4$$

$$\theta = \frac{7\pi}{6}$$

$$z = \rho \cos \phi = 8 \cos \frac{\pi}{6} = \frac{8\sqrt{3}}{2} = 4\sqrt{3}$$

Cylindrical coordinates: $\left(4, \frac{7\pi}{6}, 4\sqrt{3}\right)$

**65.** Cylindrical coordinates: $(r, \theta, z) = \left(5, \frac{3\pi}{4}, -5\right)$

First we convert to rectangular coordinates.

$$x = r \cos \theta = 5 \cos \frac{3\pi}{4} \approx -3.536$$

$$y = r \sin \theta = 5 \sin \frac{3\pi}{4} \approx 3.536$$

$$z = -5$$

Rectangular coordinates: $(-3.536, 3.536, -5)$

Converting from cylindrical coordinates to spherical coordinates, we have the following.

$$\rho = \sqrt{r^2 + z^2} = \sqrt{5^2 + (-5)^2} = 5\sqrt{2} \approx 7.071$$

$$\theta = \frac{3\pi}{4} \approx 2.356$$

$$\phi = \arccos\left(\frac{z}{\sqrt{r^2 + z^2}}\right)$$

$$= \arccos\left(\frac{-5}{5\sqrt{2}}\right) = \arccos\left(-\frac{1}{\sqrt{2}}\right) = \frac{3\pi}{4} \approx 2.356$$

Spherical coordinates: $(7.071, 2.356, 2.356)$

**79.** Rectangular coordinates: $x^2 + y^2 = 4y$

(a) Since $x^2 + y^2 = r^2$ and $y = 4 \sin \theta$, the equation in cylindrical coordinates is

$$x^2 + y^2 = 4y$$

$$r^2 = 4r \sin \theta \implies r = 4 \sin \theta.$$

(b) In spherical coordinates, since $x^2 + y^2 = r^2 = \rho^2 \sin^2 \phi$ and $y = \rho \sin \phi \sin \theta$, we have

$$x^2 + y^2 = 4y$$

$$\rho^2 \sin^2 \phi = 4\rho \sin \phi \sin \theta$$

$$\rho \sin \phi = 4 \sin \theta$$

$$\rho = 4 \sin \theta \csc \phi.$$

**87.** Spherical coordinates: $0 \le \theta \le 2\pi$, $0 \le \phi \le \pi/6$, $0 \le \rho \le a \sec \phi$

From the constraint $0 \le \rho \le a \sec \phi$, we have

$$0 \le \rho \le a\left(\frac{1}{\cos \phi}\right)$$

$$0 \le \rho \cos \phi \le a$$

$$0 \le z \le a$$

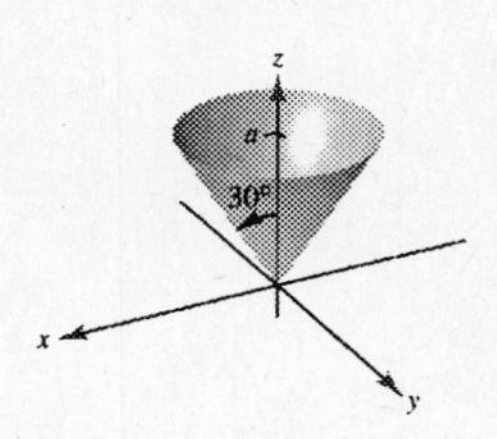

# Review Exercises for Chapter 10

**9.**
$$x^2 + y^2 + z^2 - 4x - 6y + 4 = 0$$
$$(x^2 - 4x + \_) + (y^2 - 6y + \_) + z^2 = -4 + \_ + \_$$
$$(x^2 - 4x + 4) + (y^2 - 6y + 9) + z^2 = -4 + 4 + 9$$
$$(x - 2)^2 + (y - 3)^2 + (z - 0)^2 = 9$$

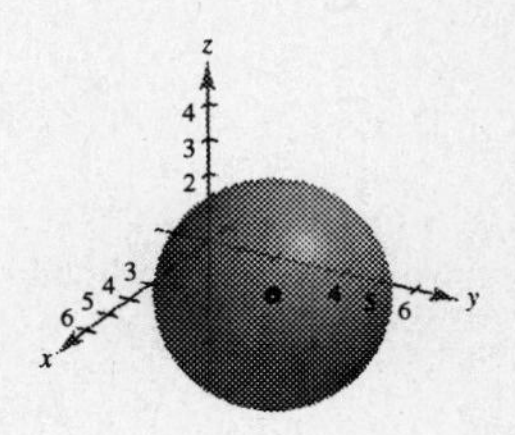

The last equation is the standard form of the equation of a sphere with center $(2, 3, 0)$ and radius 3. The graph of the sphere is given in the figure.

**11.** $P(5, 0, 0)$, $Q(4, 4, 0)$, $R(2, 0, 6)$

(a) $\mathbf{u} = \overrightarrow{PQ} = \langle 4 - 5, 4 - 0, 0 - 0\rangle = \langle -1, 4, 0\rangle$

$\mathbf{v} = \overrightarrow{PR} = \langle 2 - 5, 0 - 0, 6 - 0\rangle = \langle -3, 0, 6\rangle$

(b) $\mathbf{u} \cdot \mathbf{v} = (-1)(-3) + (4)(0) + (0)(6) = 3$

(c) $$\mathbf{u} \times \mathbf{v} = \begin{vmatrix} \mathbf{i} & \mathbf{j} & \mathbf{k} \\ -1 & 4 & 0 \\ -3 & 0 & 6 \end{vmatrix} = 24\mathbf{i} + 6\mathbf{j} + 12\mathbf{k}$$
$$= 6(4\mathbf{i} + \mathbf{j} + 2\mathbf{k})$$

(d) A vector normal to the plane is

$\frac{1}{6}(\mathbf{u} \times \mathbf{v}) = 4\mathbf{i} + \mathbf{j} + 2\mathbf{k}$ See part (c)

Therefore, using $(5, 0, 0)$ as a point in the plane, an equation of the plane is

$$4(x - 5) + 1(y - 0) + 2(z - 0) = 0$$
$$4x + y + 2z = 20.$$

(e) Since the direction of the line is determined by $\mathbf{u} = \langle -1, 4, 0\rangle$ [see part (a)], a set of parametric equations of the line passing through the point $(4, 4, 0)$ is

$$x = 4 - t,\ y = 4 + 4t,\ z = 0.$$

Note that when $t = -1$, $x = 5$, $y = 0$, and $z = 0$. Thus, the line passes through the point $P$.

**17.** $\mathbf{u} = \langle 10, -5, 15\rangle$, $\mathbf{v} = \langle -2, 1, -3\rangle$

Since $\mathbf{u} = -5\mathbf{v}$, $\mathbf{u}$ is parallel to $\mathbf{v}$ and in the opposite direction. Therefore, $\theta = \pi$. A second method for solving the problem is the following.

$$\cos\theta = \frac{\mathbf{u} \cdot \mathbf{v}}{\|\mathbf{u}\|\|\mathbf{v}\|} = \frac{-70}{5\sqrt{14}\sqrt{14}} = -1$$

Since $\cos\theta = -1$, $\theta = \arccos(-1) = \pi$.

**27.** $\mathbf{u} = \langle 3, -2, 1\rangle$, $\mathbf{w} = \langle -1, 2, 2\rangle$

$$\text{proj}_{\mathbf{u}}\mathbf{w} = \left(\frac{\mathbf{u} \cdot \mathbf{w}}{\|\mathbf{u}\|^2}\right)\mathbf{u} = -\frac{5}{14}\langle 3, -2, 1\rangle$$
$$= \left\langle -\frac{15}{14}, \frac{10}{14}, -\frac{5}{14}\right\rangle$$

**37.** (a) Any line perpendicular to the $xz$-plane must have the direction of the vector $\mathbf{v} = 0\mathbf{i} + \mathbf{j} + 0\mathbf{k}$. Thus, direction numbers for the required line are 0, 1, 0. Since the line passes through the point $(1, 2, 3)$, the parametric equations are

$$x = 1 + (0)t = 1$$
$$y = 2 + t$$
$$z = 3 + (0)t = 3.$$

(b) Since two of the direction numbers are zero, there is no symmetric form.

**41.** $\dfrac{x - 1}{-2} = y = z + 1$, $\dfrac{x + 1}{-2} = y - 1 = z - 2$

We first observe that the lines are parallel since they have the same direction numbers, $-2, 1, 1$. Therefore, a vector parallel to the plane is $\mathbf{u} = \langle -2, 1, 1\rangle$. A point on the first line is $(1, 0, -1)$ and a point on the second line is $(-1, 1, 2)$. The vector $\mathbf{v} = \langle 2, -1, -3\rangle$ connecting these two points is also parallel to the plane. Thus, a normal vector to the plane is

$$\mathbf{u} \times \mathbf{v} = \begin{vmatrix} \mathbf{i} & \mathbf{j} & \mathbf{k} \\ -2 & 1 & 1 \\ 2 & -1 & -3 \end{vmatrix} = -2(\mathbf{i} + 2\mathbf{j}).$$

Therefore, an equation of the plane is

$$1(x - 1) + 2(y - 0) + 0(z + 1) = 0$$
$$x + 2y = 1.$$

**45.** Line 1: $\dfrac{x}{1} = \dfrac{y}{2} = \dfrac{z}{3}$

Line 2: $\dfrac{x+1}{-1} = \dfrac{y}{3} = \dfrac{z+2}{2}$

Direction vectors for the two lines: $\mathbf{u} = \langle 1, 2, 3 \rangle$, $\mathbf{v} = \langle -1, 3, 2 \rangle$

A vector perpendicular to the two lines: $\mathbf{u} \times \mathbf{v} = \begin{vmatrix} \mathbf{i} & \mathbf{j} & \mathbf{k} \\ 1 & 2 & 3 \\ -1 & 3 & 2 \end{vmatrix} = \langle -5, -5, 5 \rangle$

Point on Line 1: $(1, 2, 3)$

Point on Line 2: $(0, -3, -4)$

The vector connecting the two points: $\mathbf{w} = \langle -1, -5, -7 \rangle$

$$D = \frac{|\mathbf{w} \cdot (\mathbf{u} \times \mathbf{v})|}{\|\mathbf{u} \times \mathbf{v}\|} = \frac{5}{5\sqrt{3}} = \frac{\sqrt{3}}{3}$$

**55.** (a) $z = \dfrac{1}{2}y^2 + 1 \quad (0 \le y \le 2)$

$2z = y^2 + 2$

$y = \sqrt{2z - 2} = \sqrt{2(z-1)}$

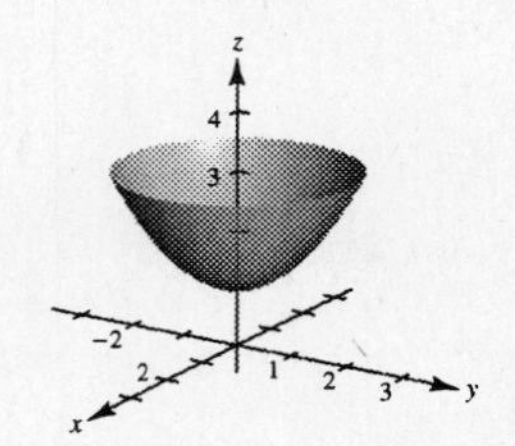


The equation of the surface generated by revolving the curve about the $z$-axis is given by

$$x^2 + y^2 = [r(z)]^2 = \left[\sqrt{2(z-1)}\right]^2$$

$$x^2 + y^2 - 2z + 2 = 0.$$

(b) $V = 2\pi \displaystyle\int_0^2 x\left[3 - \left(\frac{1}{2}x^2 + 1\right)\right] dx$

$= 2\pi \displaystyle\int_0^2 \left(2x - \frac{1}{2}x^3\right) dx$

$= 2\pi \left[x^2 - \dfrac{x^4}{8}\right]_0^2$

$= 4\pi \approx 12.6$ cubic centimeters

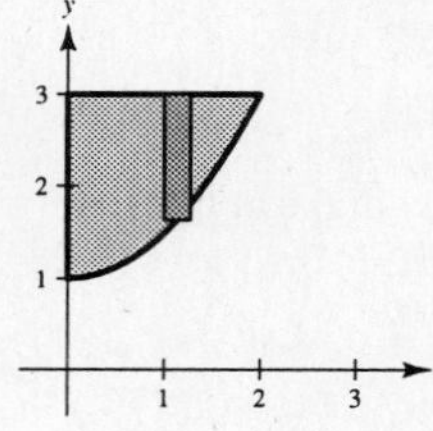


(c) $V = 2\pi \displaystyle\int_{1/2}^2 x\left[3 - \left(\frac{1}{2}x^2 + 1\right)\right] dx$

$= 2\pi \displaystyle\int_{1/2}^2 \left(2x - \frac{1}{2}x^3\right) dx$

$= 2\pi \left[x^2 - \dfrac{x^4}{8}\right]_{1/2}^2$

$= 4\pi - \dfrac{31\pi}{64} = \dfrac{225\pi}{64} \approx 11.0$ cubic centimeters

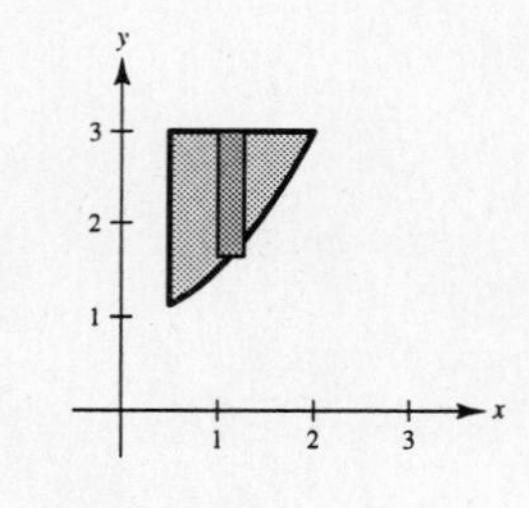

# CHAPTER 11
# Vector-Valued Functions

# CHAPTER 11
# Vector-Valued Functions

## Section 11.1 Vector-Valued Functions

**Solutions to Selected Odd-Numbered Exercises**

**3.** $\mathbf{r}(t) = \ln t\mathbf{i} - e^t\mathbf{j} - t\mathbf{k}$.

The component functions of the vector-valued function

$$\mathbf{r}(t) = f(t)\mathbf{i} + g(t)\mathbf{j} + h(t)\mathbf{k}$$

are the real-valued functions $f$, $g$, and $h$. Given the vector-valued function $\mathbf{r}(t) = \ln t\mathbf{i} - e^t\mathbf{j} - t\mathbf{k}$ the component functions and their domains are:

$$f(t) = \ln t, \qquad 0 < t < \infty$$

$$g(t) = -e^t, \qquad -\infty < t < \infty$$

$$h(t) = -t, \qquad -\infty < t < \infty.$$

The intersection of the domains of $f$, $g$, and $h$, is the interval $(0, \infty)$, the domain of $\mathbf{r}$.

**13.** $\mathbf{r}(t) = \sin 3t\mathbf{i} + \cos 3t\mathbf{j} + t\mathbf{k}$

$$\|\mathbf{r}(t)\| = \sqrt{(\sin 3t)^2 + (\cos 3t)^2 + t^2} = \sqrt{(\sin^2 3t + \cos^2 3t) + t^2} = \sqrt{1 + t^2}$$

**27.** $\mathbf{r}(t) = 2\cos t\mathbf{i} + 2\sin t\mathbf{j} + t\mathbf{k}$

From the first two parametric equations $x = 2\cos t$ and $y = 2\sin t$, we obtain

$$x^2 + y^2 = 4\cos^2 t + 4\sin^2 t = 4.$$

This means that the curve lies on a right circular cylinder of radius 2 centered about the $z$-axis. To locate the curve on this cylinder, use the third parametric equation $z = t$. Thus the graph of the vector-valued function spirals counterclockwise up the cylinder to produce a circular helix.

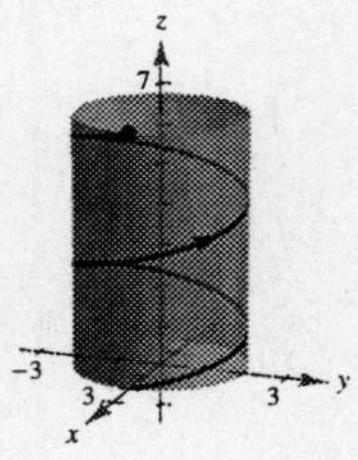


**31.** $\mathbf{r}(t) = \left\langle t, t^2, \frac{2}{3}t^3 \right\rangle$

$x = t, y = t^2, z = \frac{2}{3}t^3$

From the first two parametric equations, we obtain $y = x^2$. This means that the space curve lies on the cylinder $y = x^2$ and the vertical position of each point is determined by $z = \frac{2}{3}x^3$.

| $t$ | $-2$ | $-1$ | $0$ | $1$ | $2$ |
|---|---|---|---|---|---|
| $x$ | $-2$ | $-1$ | $0$ | $1$ | $2$ |
| $y$ | $4$ | $1$ | $0$ | $1$ | $4$ |
| $z$ | $-\frac{16}{3}$ | $-\frac{2}{3}$ | $0$ | $\frac{2}{3}$ | $\frac{16}{3}$ |

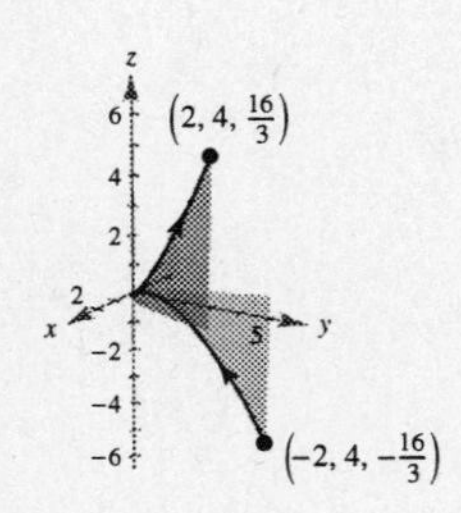

**41.** The graph of the equation $x^2 + y^2 = 25$ is a circle with center at the origin and radius 5. Let $(x, y)$ be any point on the circumference of the circle (see figure). Then

$$\cos\theta = \frac{x}{5} \Rightarrow x = 5\cos\theta$$

$$\sin\theta = \frac{y}{5} \Rightarrow y = 5\sin\theta$$

and

$$\mathbf{r}(\theta) = (5\cos\theta)\mathbf{i} + (5\sin\theta)\mathbf{j}.$$

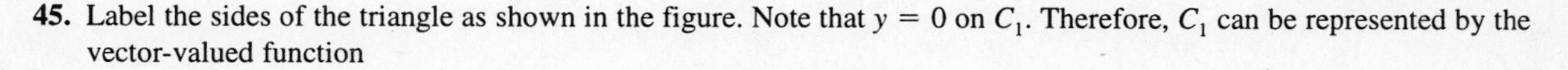

**45.** Label the sides of the triangle as shown in the figure. Note that $y = 0$ on $C_1$. Therefore, $C_1$ can be represented by the vector-valued function

$$\mathbf{r}_1(t) = t\mathbf{i},\ 0 \le t \le 4.$$

The line segment $C_2$ has slope $m = -\frac{3}{2}$ and $y$-intercept $(0, 6)$. Hence, the rectangular form of the equation is

$$y = -\tfrac{3}{2}x + 6.$$

Letting $x = 2t$ and substituting this expression for $x$ in the rectangular equation of the line yields $y = -3t + 6$. However, for increasing values of the parameter, $x$ increases and $y$ decreases which is the incorrect orientation. Replacing $t$ by $2 - t$ reverses the orientation and produces

$$x = 2(2 - t) = 4 - 2t \text{ and } y = -3(2 - t) + 6 = 3t.$$

Therefore,

$$\mathbf{r}_2(t) = (4 - 2t)\mathbf{i} + 3t\mathbf{j},\ 0 \le t \le 2.$$

Note that $x = 0$ on $C_3$. Since $y$ is decreasing on $C_3$, a vector-valued function for $C_3$ is

$$\mathbf{r}_3(t) = (6 - t)\mathbf{j},\ 0 \le t \le 6.$$

(Observe that there are many correct answers for this problem.)

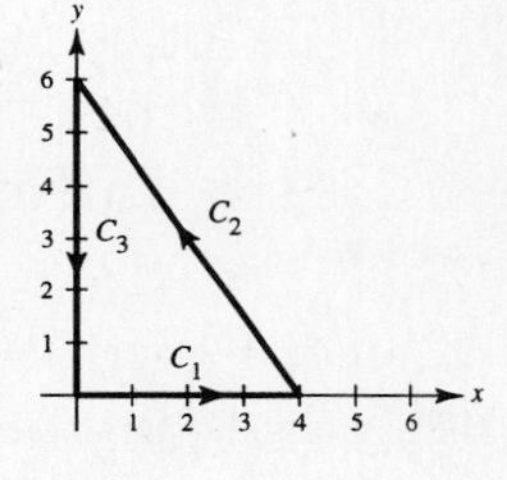


**53.** The equation $x^2 + y^2 + z^2 = 4$ represents a sphere of radius 2 centered at the origin. The equation $x + z = 2$ represents a plane. If we let $x = 1 + \sin t$, then

$$z = 2 - x = 1 - \sin t.$$

Substituting into the equation of the sphere, we have

$$\begin{aligned}
x^2 + y^2 + z^2 &= 4\\
(1 + \sin t)^2 + y^2 + (1 - \sin t)^2 &= 4\\
2 + 2\sin^2 t + y^2 &= 4\\
y^2 &= 2 - 2\sin^2 t\\
y^2 &= 2\cos^2 t\\
y &= \pm\sqrt{2}\cos t.
\end{aligned}$$

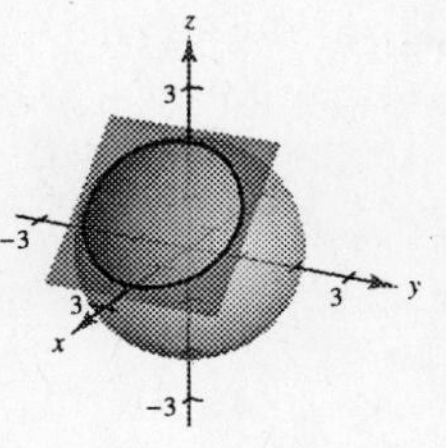


Therefore, $x = 1 + \sin t$, $y = \pm\sqrt{2}\cos t$, $z = 1 - \sin t$ and the two vector-valued functions are

$$\mathbf{r}(t) = (1 + \sin t)\mathbf{i} + \left(\sqrt{2}\cos t\right)\mathbf{j} + (1 - \sin t)\mathbf{k}$$

and

$$\mathbf{r}(t) = (1 + \sin t)\mathbf{i} - \left(\sqrt{2}\cos t\right)\mathbf{j} + (1 - \sin t)\mathbf{k}.$$

| $t$ | $\frac{-\pi}{2}$ | $\frac{-\pi}{6}$ | $0$ | $\frac{\pi}{6}$ | $\frac{\pi}{2}$ |
|---|---|---|---|---|---|
| $x$ | $0$ | $\frac{1}{2}$ | $1$ | $\frac{3}{2}$ | $2$ |
| $y$ | $0$ | $\pm\frac{\sqrt{6}}{2}$ | $\pm\sqrt{2}$ | $\pm\frac{\sqrt{6}}{2}$ | $0$ |
| $z$ | $2$ | $\frac{3}{2}$ | $1$ | $\frac{1}{2}$ | $0$ |

**55.** The equation $x^2 + z^2 = 4$ represents a cylinder of radius 2 with its axis along the $y$-axis, while the equation $y^2 + z^2 = 4$ represents a cylinder of radius 2 with its axis along the $x$-axis. Thus the curve of intersection lies along both cylinders, as shown in the sketch. Two points on the curve are $(2, 2, 0)$ and $(0, 0, 2)$. To find a set of parametric equations for the curve of intersection, subtract the second equation from the first to obtain

$$\begin{aligned} x^2 + z^2 &= 4 \\ -(y^2 + z^2 &= 4) \\ \hline x^2 - y^2 &= 0 \text{ or } y = \pm x. \end{aligned}$$

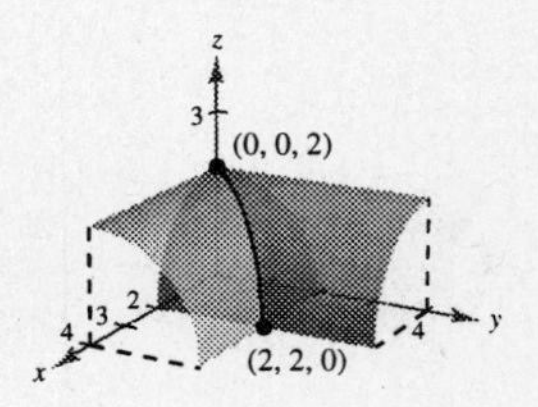


Therefore, in the first octant, if we let $x = t(t > 0)$, then parametric equations for the curve are

$$x = t,\ y = t,\ z = \sqrt{4 - t^2}$$

and the vector-valued function is

$$\mathbf{r}(t) = t\mathbf{i} + t\mathbf{j} + \sqrt{4 - t^2}\mathbf{k}.$$

**59.** $$\lim_{t\to 0}\left(t^2\mathbf{i} + 3t\mathbf{j} + \frac{1 - \cos t}{t}\mathbf{k}\right) = \left[\lim_{t\to 0} t^2\right]\mathbf{i} + \left[\lim_{t\to 0} 3t\right]\mathbf{j} + \left[\lim_{t\to 0}\frac{1 - \cos t}{t}\right]\mathbf{k}$$

$$= 0\mathbf{i} + 0\mathbf{j} + 0\mathbf{k} = \mathbf{0}$$

**65.** $\mathbf{r}(t) = t\mathbf{i} + \arcsin t\mathbf{j} + (t - 1)\mathbf{k}$

is continuous on $[-1, 1]$. This interval is the domain of $\mathbf{r}$ since it is the domain of the arcsine function.

## Section 11.2 Differentiation and Integration of Vector-Valued Functions

**3.** $\mathbf{r}(t) = \cos t\mathbf{i} + \sin t\mathbf{j}$

(a) Eliminating the parameter from the parametric equations $x = \cos t$ and $y = \sin t$, we have

$$x^2 + y^2 = \cos^2 t + \sin^2 t = 1.$$

Therefore, the vector-valued function represents a circle of radius 1 center at the origin. (See the figure.)

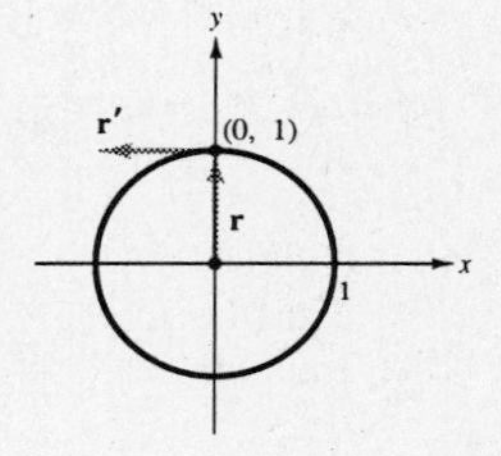


(b) $\mathbf{r}(t) = \cos t\mathbf{i} + \sin t\mathbf{j} \quad \Rightarrow \mathbf{r}\left(\frac{\pi}{2}\right) = \mathbf{j}$

$\mathbf{r}'(t) = -\sin t\mathbf{i} + \cos t\mathbf{j} \Rightarrow \mathbf{r}'\left(\frac{\pi}{2}\right) = -\mathbf{i}$

**11.** $\mathbf{r}(t) = 6t\mathbf{i} - 7t^2\mathbf{j} + t^3\mathbf{k}$

$$\mathbf{r}'(t) = \frac{d}{dt}[6t]\mathbf{i} + \frac{d}{dt}[-7t^2]\mathbf{j} + \frac{d}{dt}[t^3]\mathbf{k}$$

$$= 6\mathbf{i} - 14t\mathbf{j} + 3t^2\mathbf{k}$$

**17.** $\mathbf{r}(t) = \langle t\sin t, t\cos t, t\rangle$

$$\mathbf{r}'(t) = \left\langle\frac{d}{dt}[t\sin t], \frac{d}{dt}[t\cos t], \frac{d}{dt}[t]\right\rangle$$

$$= \langle t\cos t + \sin t, -t\sin t + \cos t, 1\rangle$$

**19.** $\mathbf{r}(t) = t\mathbf{i} + 3t\mathbf{j} + t^2\mathbf{k}$, $\mathbf{u}(t) = 4t\mathbf{i} + t^2\mathbf{j} + t^3\mathbf{k}$

(a) Differentiating in a component-by-component basis produces

$$\mathbf{r}'(t) = \mathbf{i} + 3\mathbf{j} + 2t\mathbf{k}.$$

(b) Differentiating $\mathbf{r}'(t)$ in a component-by-component basis produces

$$\mathbf{r}''(t) = 2\mathbf{k}.$$

(c) $$\mathbf{r}(t)\cdot\mathbf{u}(t) = t(4t) + 3t(t^2) + t^2(t^3)$$

$$= 4t^2 + 3t^3 + t^5$$

$$D_t[\mathbf{r}(t)\cdot\mathbf{u}(t)] = 8t + 9t^2 + 5t^4$$

(d) $$3\mathbf{r}(t) - \mathbf{u}(t) = 3(t\mathbf{i} + 3t\mathbf{j} + t^2\mathbf{k}) - (4t\mathbf{i} + t^2\mathbf{j} + t^3\mathbf{k})$$

$$= -t\mathbf{i} + (9t - t^2)\mathbf{j} + (3t^2 - t^3)\mathbf{k}$$

$$D_t[3\mathbf{r}(t) - \mathbf{u}(t)] = -\mathbf{i} + (9 - 2t)\mathbf{j} + (6t - 3t^2)\mathbf{k}$$

—CONTINUED—

**19. —CONTINUED—**

(e) $$\mathbf{r}(t) \times \mathbf{u}(t) = \begin{vmatrix} \mathbf{i} & \mathbf{j} & \mathbf{k} \\ t & 3t & t^2 \\ 4t & t^2 & t^3 \end{vmatrix}$$

$$= \begin{vmatrix} 3t & t^2 \\ t^2 & t^3 \end{vmatrix}\mathbf{i} - \begin{vmatrix} t & t^2 \\ 4t & t^3 \end{vmatrix}\mathbf{j} + \begin{vmatrix} t & 3t \\ 4t & t^2 \end{vmatrix}\mathbf{k}$$

$$= 2t^4\mathbf{i} - (t^4 - 4t^3)\mathbf{j} + (t^3 - 12t^2)\mathbf{k}$$

$$D_t[\mathbf{r}(t) \times \mathbf{u}(t)] = 8t^3\mathbf{i} + (12t^2 - 4t^3)\mathbf{j} + (3t^2 - 24t)\mathbf{k}$$

(f) $$\|r(t)\| = \sqrt{t^2 + (3t)^2 + (t^2)^2}$$

$$= \sqrt{10t^2 + t^4}$$

$$D_t[\|\mathbf{r}(t)\|] = \frac{1}{2}(10t^2 + t^4)^{-1/2}(20t + 4t^3)$$

$$= \frac{10t + 2t^3}{\sqrt{10t^2 + t^4}} = \frac{10 + 2t^2}{\sqrt{10 + t^2}}$$

**29.** $\mathbf{r}(t) = (t - 1)\mathbf{i} + \dfrac{1}{t}\mathbf{j} - t^2\mathbf{k}$

$\mathbf{r}'(t) = \mathbf{i} - \dfrac{1}{t^2}\mathbf{j} - 2t\mathbf{k}$

$\mathbf{r}$ is not continuous at $t = 0$ and $\mathbf{r}'(t) \neq 0$ for all $t$ in the domain of $\mathbf{r}$. Therefore, the curve is smooth on $(-\infty, 0)$ and $(0, \infty)$.

**39.** $$\int\left(\frac{1}{t}\mathbf{i} + \mathbf{j} - t^{3/2}\mathbf{k}\right)dt = \left[\int \frac{1}{t}\,dt\right]\mathbf{i} + \left[\int dt\right]\mathbf{j} + \left[\int -t^{3/2}\,dt\right]\mathbf{k}$$

$$= \ln|t|\mathbf{i} + t\mathbf{j} - \frac{2}{5}t^{5/2}\mathbf{k} + \mathbf{C}$$

**45.** $\mathbf{r}'(t) = 4e^{2t}\mathbf{i} + 3e^t\mathbf{j}$, $\mathbf{r}(0) = 2\mathbf{i}$

$$\mathbf{r}(t) = \mathbf{i}\int 4e^{2t}\,dt + \mathbf{j}\int 3e^t\,dt = \mathbf{i}[2e^{2t} + c_1] + \mathbf{j}[3e^t + c_2]$$

$$\mathbf{r}(0) = \mathbf{i}[2 + c_1] + \mathbf{j}[3 + c_2] = 2\mathbf{i}$$

Therefore, $2 + c_1 = 2$, or $c_1 = 0$. Also, $3 + c_2 = 0$, or $c_2 = -3$. Thus,

$$\mathbf{r}(t) = 2e^{2t}\mathbf{i} + (3e^t - 3)\mathbf{j} = 2e^{2t}\mathbf{i} + 3(e^t - 1)\mathbf{j}.$$

**51.** $$\int_0^1 (8t\mathbf{i} + t\mathbf{j} - \mathbf{k})\,dt = \mathbf{i}\int_0^1 8t\,dt + \mathbf{j}\int_0^1 t\,dt - \mathbf{k}\int_0^1 dt$$

$$= \Big[4t^2\Big]_0^1\mathbf{i} + \left[\frac{1}{2}t^2\right]_0^1\mathbf{j} - \Big[t\Big]_0^1\mathbf{k}$$

$$= 4\mathbf{i} + \frac{1}{2}\mathbf{j} - \mathbf{k}$$

## Section 11.3 Velocity and Acceleration

**3.** $\mathbf{r}(t) = t^2\mathbf{i} + t\mathbf{j}$

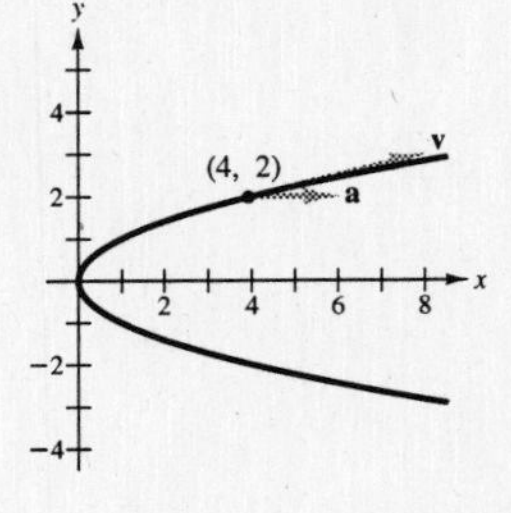


Eliminating the parameter from the parametric equations $x = t^2$ and $y = t$, we obtain the rectangular equation $x = y^2$. Therefore, the object is moving in a parabolic path. (See the figure.) The velocity is given by

$$\mathbf{v}(t) = \mathbf{r}'(t) = 2t\mathbf{i} + \mathbf{j}$$

and the acceleration is given by

$$\mathbf{a}(t) = \mathbf{r}''(t) = 2\mathbf{i}.$$

At the point $(4, 2)$, $t = 2$. Thus,

$$\mathbf{v}(2) = 4\mathbf{i} + \mathbf{j} \quad \text{and} \quad \mathbf{a}(2) = 2\mathbf{i}.$$

**15.** Since

$$\mathbf{r}(t) = \langle 4t, 3\cos t, 3\sin t\rangle,$$

we have

$$\mathbf{v}(t) = \mathbf{r}'(t) = \langle 4, -3\sin t, 3\cos t\rangle,$$

$$\text{speed} = \|\mathbf{v}(t)\| = \sqrt{16 + 9(\sin^2 t + \cos^2 t)} = \sqrt{25} = 5,$$

and

$$\mathbf{a}(t) = \mathbf{r}''(t) = \langle 0, -3\cos t, -3\sin t\rangle.$$

**21.** $\mathbf{a}(t) = t\mathbf{j} + t\mathbf{k}$, $\mathbf{v}(1) = 5\mathbf{j}$, $\mathbf{r}(1) = \mathbf{0}$

$$\mathbf{v}(t) = \int \mathbf{a}(t)\,dt + \mathbf{C} = \int (t\mathbf{j} + t\mathbf{k})\,dt + \mathbf{C} = \frac{1}{2}t^2\mathbf{j} + \frac{1}{2}t^2\mathbf{k} + C_1\mathbf{i} + C_2\mathbf{j} + C_3\mathbf{k}$$

$$\mathbf{v}(1) = C_1\mathbf{i} + \left(\frac{1}{2} + C_2\right)\mathbf{j} + \left(\frac{1}{2} + C_3\right)\mathbf{k} = 5\mathbf{j}$$

Therefore,

$$C_1 = 0$$

$$\frac{1}{2} + C_2 = 5 \Rightarrow C_2 = \frac{9}{2}$$

$$\frac{1}{2} + C_3 = 0 \Rightarrow C_3 = -\frac{1}{2}.$$

Thus, the velocity vector is

$$\mathbf{v}(t) = \left(\frac{t^2}{2} + \frac{9}{2}\right)\mathbf{j} + \left(\frac{t^2}{2} - \frac{1}{2}\right)\mathbf{k}$$

$$\mathbf{r}(t) = \int \mathbf{v}(t)\,dt + \mathbf{C} = \int\left(\frac{t^2}{2} + \frac{9}{2}t\right)dt\,\mathbf{j} + \int\left(\frac{t^2}{2} - \frac{1}{2}\right)dt\,\mathbf{k} + \mathbf{C} = \left(\frac{t^3}{6} + \frac{9}{2}t\right)\mathbf{j} + \left(\frac{t^3}{6} - \frac{1}{2}t\right)\mathbf{k} + C_4\mathbf{i} + C_5\mathbf{j} + C_6\mathbf{k}$$

$$\mathbf{r}(1) = C_4\mathbf{i} + \left(\frac{1}{6} + \frac{9}{2} + C_5\right)\mathbf{j} + \left(\frac{1}{6} - \frac{1}{2} + C_6\right)\mathbf{k} = C_4\mathbf{i} + \left(\frac{14}{3} + C_5\right)\mathbf{j} + \left(-\frac{1}{3} + C_6\right)\mathbf{k} = \mathbf{0}.$$

This implies that

$$C_4 = 0,\ C_5 = \frac{-14}{3}, \text{ and } C_6 = \frac{1}{3}.$$

Thus, the position vector is

$$\mathbf{r}(t) = \left(\frac{t^3}{6} + \frac{9}{2}t - \frac{14}{3}\right)\mathbf{j} + \left(\frac{t^3}{6} - \frac{1}{2}t + \frac{1}{3}\right)\mathbf{k} \quad \text{and} \quad \mathbf{r}(2) = \frac{17}{3}\mathbf{j} + \frac{2}{3}\mathbf{k}.$$

**25.** The path of the ball is given by

$$\mathbf{r}(t) = (v_0 \cos 45°)t\mathbf{i} + [3 + (v_0 \sin 45°)t - 16t^2]\mathbf{j}$$
$$= \left(\frac{tv_0}{\sqrt{2}}\right)\mathbf{i} + \left(3 + \frac{tv_0}{\sqrt{2}} - 16t^2\right)\mathbf{j}.$$

We know that the horizontal component is 300 when the vertical component is three. Thus,

$$\frac{tv_0}{\sqrt{2}} = 300 \quad \text{and} \quad 3 + \frac{tv_0}{\sqrt{2}} - 16t^2 = 3.$$

From the first equation we obtain $t = 300\sqrt{2}/v_0$. Substituting this expression into the second equation yields

$$\frac{300\sqrt{2}}{v_0}\left(\frac{v_0}{\sqrt{2}}\right) - 16\left(\frac{300\sqrt{2}}{v_0}\right)^2 = 0$$
$$300 = \frac{16(300^2)(2)}{v_0^2}$$
$$v_0^2 = 32(300)$$
$$v_0 = \sqrt{9600} = 40\sqrt{6} \approx 97.98 \text{ ft/sec.}$$

The maximum height is reached when the derivative of the vertical component is zero. Thus,

$$y(t) = 3 + \frac{tv_0}{\sqrt{2}} - 16t^2 = 3 + \frac{t(40\sqrt{6})}{\sqrt{2}} - 16t^2 = 3 + 40\sqrt{3}t - 16t^2$$
$$y'(t) = 40\sqrt{3} - 32t = 0$$
$$t = \frac{40\sqrt{3}}{32} = \frac{5\sqrt{3}}{4}.$$

Finally, the maximum height is

$$y\left(\frac{5\sqrt{3}}{4}\right) = 3 + 40\sqrt{3}\left(\frac{5\sqrt{3}}{4}\right) - 16\left(\frac{5\sqrt{3}}{4}\right)^2$$
$$= 3 + 150 - 75 = 78 \text{ ft.}$$

**29.** If we place the origin of the coordinate system at the ejector on the baler (see figure), then

$$\mathbf{r}(t) = (v \cos \theta)t\mathbf{i} + [(v \sin \theta)t - 16t^2]\mathbf{j}.$$

Since the bale must be thrown to the position (16, 8), we have

$$16 = (v \cos \theta)t \implies t = \frac{16}{v \cos \theta}$$
$$8 = (v \sin \theta)t - 16t^2.$$



Substituting into the second equation and solving for $v$, we obtain the following.

$$8 = (v \sin \theta)\left(\frac{16}{v \cos \theta}\right) - 16\left(\frac{16}{v \cos \theta}\right)^2$$
$$1 = \frac{2 \sin \theta}{\cos \theta} - \frac{512}{v^2 \cos^2 \theta}$$
$$\frac{512}{v^2 \cos^2 \theta} = \frac{2 \sin \theta}{\cos \theta} - 1$$
$$\frac{1}{v^2} = \left(\frac{2 \sin \theta}{\cos \theta} - 1\right)\left(\frac{\cos^2 \theta}{512}\right) = \frac{2 \sin \theta \cos \theta - \cos^2 \theta}{512}$$
$$v^2 = \frac{512}{2 \sin \theta \cos \theta - \cos^2 \theta}$$

**—CONTINUED—**

**29. —CONTINUED—**

We now must minimize $f(\theta) = \dfrac{512}{2\sin\theta\cos\theta - \cos^2\theta}$.

$$f'(\theta) = \frac{-512(2\cos^2\theta - 2\sin^2\theta + 2\sin\theta\cos\theta)}{(2\sin\theta\cos\theta - \cos^2\theta)^2}$$

$$= \frac{-512(2\cos 2\theta + \sin 2\theta)}{(2\sin\theta\cos\theta - \cos^2\theta)^2}$$

$$f'(\theta) = 0 \implies 2\cos 2\theta + \sin 2\theta = 0$$

$$\tan 2\theta = -2$$

$$\theta \approx 1.01722 \approx 58.28°$$

Substituting this into the equation for $v$ yields $v \approx 28.78$ feet per second.

**39.**

$$\mathbf{r}(t) = b(\omega t - \sin\omega t)\mathbf{i} + b(1 - \cos\omega t)\mathbf{j}$$

$$\mathbf{r}'(t) = \mathbf{v}(t) = b(\omega - \omega\cos\omega t)\mathbf{i} + b\omega\sin\omega t\,\mathbf{j}$$

$$\text{speed} = \|\mathbf{v}(t)\|$$

$$= \sqrt{[b(\omega - \omega\cos\omega t)]^2 + [b\omega\sin\omega t]^2}$$

$$= b\omega\sqrt{1 - 2\cos\omega t + \cos^2\omega t + \sin^2\omega t}$$

$$= b\omega\sqrt{2 - 2\cos\omega t} = \sqrt{2}b\omega\sqrt{1 - \cos\omega t}$$

(a) When $\omega t = 0, 2\pi, 4\pi, \dots$, $1 - \cos\omega t = 0$, and therefore, $\|\mathbf{v}(t)\| = 0$. Hence, the speed is zero when the point contacts the surface on which the circle is rolling.

(b) When $\omega t = \pi, 3\pi, \dots$, $1 - \cos\omega t = 2$, and therefore, $\|\mathbf{v}(t)\| = 2b\omega$, its maximum value. Hence, the speed is maximum when the point is at the top of cycloidal arch.

**41.** $\mathbf{r}(t) = b\cos\omega t\mathbf{i} + b\sin\omega t\mathbf{j}$

The velocity vector is

$$\mathbf{v}(t) = \mathbf{r}'(t) = -b\omega\sin\omega t\mathbf{i} + b\omega\cos\omega t\mathbf{j}$$

and since

$$\mathbf{r}(t)\cdot\mathbf{v}(t) = -b^2\omega\sin\omega t\cos\omega t + b^2\omega\sin\omega t\cos\omega t = 0,$$

it follows that $\mathbf{r}(t)$ and $\mathbf{v}(t)$ are orthogonal.

**43.** $\mathbf{r}(t) = b\cos\omega t\mathbf{i} + b\sin\omega t\mathbf{j}$

$$\mathbf{r}'(t) = -b\omega\sin\omega t\mathbf{i} + b\omega\cos\omega t\mathbf{j}$$

$$\mathbf{a}(t) = \mathbf{r}''(t) = [-b\omega^2\cos\omega t]\mathbf{i} - [b\omega^2\sin\omega t]\mathbf{j}$$

$$= -b\omega^2[\cos\omega t\mathbf{i} + \sin\omega t\mathbf{j}]$$

$$= -\omega^2\,\mathbf{r}(t)$$

Therefore, $\mathbf{a}(t)$ is a negative multiple of a unit vector from $(0, 0)$ to $(\cos\omega t, \sin\omega t)$, and thus $\mathbf{a}(t)$ is directed toward the origin.

# Section 11.4 Tangent Vectors and Normal Vectors

**3.** $\mathbf{r}(t) = 2\cos t\mathbf{i} + 2\sin t\mathbf{j} + t\mathbf{k}$

$\mathbf{r}'(t) = -2\sin t\mathbf{i} + 2\cos t\mathbf{j} + \mathbf{k}$

The unit tangent vector is

$$\mathbf{T}(t) = \frac{\mathbf{r}'(t)}{\|\mathbf{r}'(t)\|} = \frac{-2\sin t\mathbf{i} + 2\cos t\mathbf{j} + \mathbf{k}}{\sqrt{(-2\sin t)^2 + (2\cos t)^2 + 1}}$$

$$= \frac{1}{\sqrt{5}}(-2\sin t\mathbf{i} + 2\cos t\mathbf{j} + \mathbf{k}).$$

Since $t = 0$ at the point $(2, 0, 0)$, the direction vector for the line is given by $\mathbf{r}'(0) = 2\mathbf{j} + \mathbf{k}$, and the parametric representation of the line is

$x = 2, y = 2s, z = s.$

**11.** $\mathbf{r}(t) = \left\langle t - 2, t^2, \frac{1}{2}t \right\rangle, \mathbf{u}(s) = \left\langle \frac{1}{4}s, 2s, \sqrt[3]{s} \right\rangle$

The space curves intersect since

$\mathbf{r}(4) = \langle 2, 16, 2 \rangle = \mathbf{u}(8).$

Now determine the tangent vectors at the specified values of the parameters.

$$\mathbf{r}'(t) = \left\langle 1, 2t, \frac{1}{2} \right\rangle$$

$$\mathbf{r}'(4) = \left\langle 1, 2, \frac{1}{2} \right\rangle$$

$$\mathbf{u}'(s) = \left\langle \frac{1}{4}, 2, \frac{1}{3s^{2/3}} \right\rangle$$

$$\mathbf{u}'(8) = \left\langle \frac{1}{4}, 2, \frac{1}{12} \right\rangle$$

The angle $\theta$ between the tangent vectors to the curves at the point of intersection is

$$\theta = \arccos\left(\frac{\mathbf{r}'(4) \cdot \mathbf{u}'(8)}{\|\mathbf{r}'(4)\|\|\mathbf{u}'(8)\|}\right)$$

$$= \arccos\left(\frac{\frac{391}{24}}{\frac{3\sqrt{29}}{2} \cdot \frac{\sqrt{586}}{12}}\right)$$

$$\approx \arccos(0.9998) \approx 0.0206 \approx 1.2^\circ.$$

**17.** $\mathbf{r}(t) = t\mathbf{i} + \frac{1}{t}\mathbf{j}$

$$\mathbf{v}(t) = \mathbf{r}'(t) = \mathbf{i} - \frac{1}{t^2}\mathbf{j}$$

$$\mathbf{a}(t) = \mathbf{r}''(t) = \frac{2}{t^3}\mathbf{j}$$

At $t = 1$, we have

$\mathbf{v}(1) = \mathbf{i} - \mathbf{j}$, $\|\mathbf{v}(1)\| = \sqrt{2}$, and $\mathbf{a}(1) = 2\mathbf{j}$.

Therefore, when $t = 1$,

$$\mathbf{T}(1) = \frac{\mathbf{v}(1)}{\|\mathbf{v}(1)\|} = \frac{\mathbf{i}}{\sqrt{2}} - \frac{\mathbf{j}}{\sqrt{2}} = \frac{\sqrt{2}}{2}(\mathbf{i} - \mathbf{j}).$$

Since $\mathbf{N}(1)$ points toward the concave side of the curve (see figure),

$$\mathbf{N}(1) = \frac{\mathbf{i}}{\sqrt{2}} + \frac{\mathbf{j}}{\sqrt{2}} = \frac{\sqrt{2}}{2}(\mathbf{i} + \mathbf{j}).$$

It follows that

$$a_{\mathbf{T}} = \mathbf{a}(1) \cdot \mathbf{T}(1) = (2\mathbf{j}) \cdot \left(\frac{\mathbf{i}}{\sqrt{2}} - \frac{\mathbf{j}}{\sqrt{2}}\right)$$

$$= \frac{-2}{\sqrt{2}} = -\sqrt{2}$$

and

$$a_{\mathbf{N}} = \mathbf{a}(1) \cdot \mathbf{N}(1) = (2\mathbf{j}) \cdot \left(\frac{\mathbf{i}}{\sqrt{2}} + \frac{\mathbf{j}}{\sqrt{2}}\right)$$

$$= \frac{2}{\sqrt{2}} = \sqrt{2}.$$

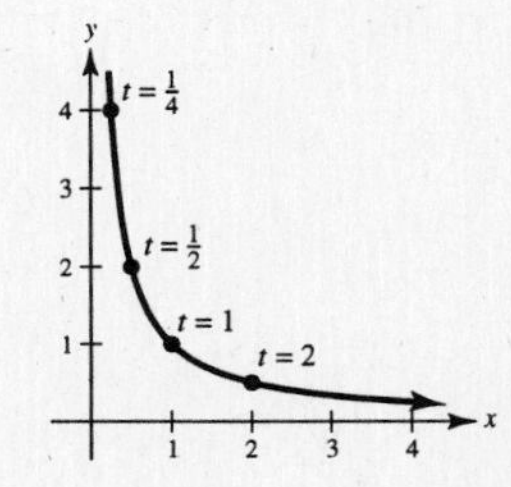

**19.** $\mathbf{r}(t) = e^t \cos t\mathbf{i} + e^t \sin t\mathbf{j}$

$\mathbf{v}(t) = \mathbf{r}'(t) = e^t(\cos t - \sin t)\mathbf{i} + e^t(\cos t + \sin t)\mathbf{j}$

$\mathbf{a}(t) = \mathbf{r}''(t) = e^t(-\sin t - \cos t + \cos t - \sin t)\mathbf{i} + e^t(-\sin t + \cos t + \cos t + \sin t)\mathbf{j}$

$= e^t(-2\sin t)\mathbf{i} + e^t(2\cos t)\mathbf{j}$

At $t = \pi/2$, we have

$$\mathbf{v}\left(\frac{\pi}{2}\right) = -e^{\pi/2}\mathbf{i} + e^{\pi/2}\mathbf{j}$$

$$\left\|\mathbf{v}\left(\frac{\pi}{2}\right)\right\| = e^{\pi/2}\sqrt{2}$$

$$\mathbf{a}\left(\frac{\pi}{2}\right) = -2e^{\pi/2}\mathbf{i}.$$

Therefore, at $t = \pi/2$,

$$\mathbf{T}\left(\frac{\pi}{2}\right) = \frac{\mathbf{v}(\pi/2)}{\|\mathbf{v}(\pi/2)\|} = \frac{-\mathbf{i}}{\sqrt{2}} + \frac{\mathbf{j}}{\sqrt{2}} = \frac{\sqrt{2}}{2}(-\mathbf{i} + \mathbf{j}),$$

and since $\mathbf{N}(\pi/2)$ points toward the concave side of the curve, (see figure)

$$\mathbf{N}\left(\frac{\pi}{2}\right) = -\frac{\sqrt{2}}{2}(\mathbf{i} + \mathbf{j}).$$

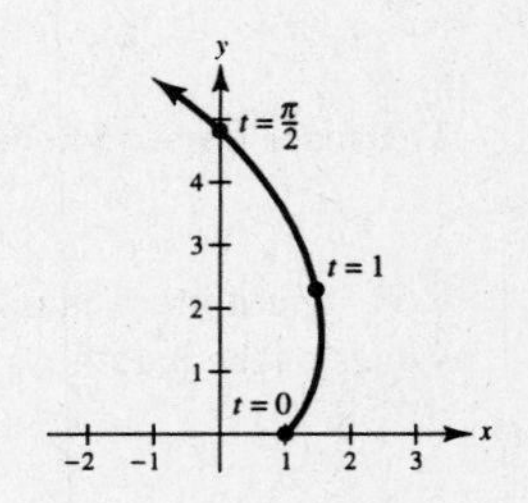


It follows that

$$a_{\mathbf{T}} = \mathbf{a}\left(\frac{\pi}{2}\right) \cdot \mathbf{T}\left(\frac{\pi}{2}\right) = (-2e^{\pi/2}\mathbf{i}) \cdot \frac{(-\mathbf{i} + \mathbf{j})}{\sqrt{2}} = \sqrt{2}e^{\pi/2}$$

$$a_{\mathbf{N}} = \mathbf{a}\left(\frac{\pi}{2}\right) \cdot \mathbf{N}\left(\frac{\pi}{2}\right) = (-2e^{\pi/2}\mathbf{i}) \cdot \frac{(-\mathbf{i} - \mathbf{j})}{\sqrt{2}} = \sqrt{2}e^{\pi/2}.$$

**29.** (a) $\mathbf{r}(t) = \langle \pi t - \sin \pi t, 1 - \cos \pi t\rangle$

$\mathbf{v}(t) = \langle \pi - \pi\cos \pi t, \pi \sin \pi t\rangle$

$\mathbf{a}(t) = \langle \pi^2 \sin \pi t, \pi^2 \cos \pi t\rangle$

$$\mathbf{T}(t) = \frac{\mathbf{v}(t)}{\|\mathbf{v}(t)\|} = \frac{1}{\sqrt{2(1 - \cos \pi t)}}\langle 1 - \cos \pi t, \sin \pi t\rangle$$

$$\mathbf{N}(t) = \frac{\mathbf{T}'(t)}{\|\mathbf{T}'(t)\|} = \frac{1}{\sqrt{2(1 - \cos \pi t)}}\langle \sin \pi t, -1 + \cos \pi t\rangle$$

$$a_{\mathbf{T}} = \mathbf{a} \cdot \mathbf{T} = \frac{1}{\sqrt{2(1 - \cos \pi t)}}[\pi^2 \sin \pi t(1 - \cos \pi t) + \pi^2 \cos \pi t \sin \pi t] = \frac{\pi^2 \sin \pi t}{\sqrt{2(1 - \cos \pi t)}}$$

$$a_{\mathbf{N}} = \mathbf{a} \cdot \mathbf{N} = \frac{1}{\sqrt{2(1 - \cos \pi t)}}[\pi^2 \sin^2 \pi t + \pi^2 \cos \pi t(-1 + \cos \pi t)] = \frac{\pi^2(1 - \cos \pi t)}{\sqrt{2(1 - \cos \pi t)}} = \frac{\pi^2\sqrt{2(1 - \cos \pi t)}}{2}$$

When $t = \frac{1}{2}$: $a_{\mathbf{T}} = \frac{\pi^2}{\sqrt{2}} = \frac{\sqrt{2}\pi^2}{2}$, $a_{\mathbf{N}} = \frac{\sqrt{2}\pi^2}{2}$

When $t = 1$: $a_{\mathbf{T}} = 0$, $a_{\mathbf{N}} = \pi^2$

When $t = \frac{3}{2}$: $a_{\mathbf{T}} = -\frac{\sqrt{2}\pi^2}{2}$, $a_{\mathbf{N}} = \frac{\sqrt{2}\pi^2}{2}$

—CONTINUED—

**29. —CONTINUED—**

(b) Speed: $s = \|\mathbf{v}(t)\| = \pi\sqrt{2(1 - \cos \pi t)}$

$$\frac{ds}{dt} = \frac{\pi^2 \sin \pi t}{\sqrt{2(1 - \cos \pi t)}} = a_{\mathbf{T}}$$

When $t = \frac{1}{2}$: $a_{\mathbf{T}} = \frac{\sqrt{2}\pi^2}{2} > 0 \Rightarrow$ the speed is increasing.

When $t = 1$: $a_{\mathbf{T}} = 0 \Rightarrow$ the speed is maximum.

When $t = \frac{3}{2}$: $a_{\mathbf{T}} = -\frac{\sqrt{2}\pi^2}{2} < 0 \Rightarrow$ the speed is decreasing.

When the sign of $a_{\mathbf{T}}$ and $a_{\mathbf{N}}$ are the same the speed increases, and when they have opposite signs the speed decreases.

**35.** $\mathbf{r}(t) = 4t\mathbf{i} + 3\cos t\mathbf{j} + 3\sin t\mathbf{k}$

$\mathbf{v}(t) = \mathbf{r}'(t) = 4\mathbf{i} - 3\sin t\mathbf{j} + 3\cos t\mathbf{k}$

$\|\mathbf{v}(t)\| = \sqrt{16 + 9(\sin^2 t + \cos^2 t)} = \sqrt{25} = 5$

$\mathbf{a}(t) = \mathbf{r}''(t) = -3\cos t\mathbf{j} - 3\sin t\mathbf{k}$

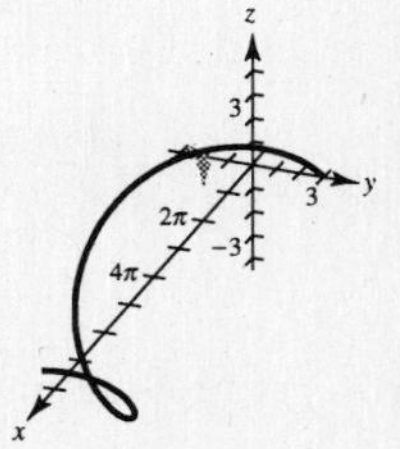

$$\mathbf{T}(t) = \frac{\mathbf{v}(t)}{\|\mathbf{v}(t)\|} = \frac{1}{5}[4\mathbf{i} - 3\sin t\mathbf{j} + 3\cos t\mathbf{k}]$$

$$\mathbf{N}(t) = \frac{\mathbf{T}'(t)}{\|\mathbf{T}'(t)\|} = \frac{(1/5)[-3\cos t\mathbf{j} - 3\sin t\mathbf{k}]}{(1/5)\sqrt{9(\cos^2 t + \sin^2 t)}} = \frac{(-3/5)[(\cos t)\mathbf{j} + (\sin t)\mathbf{k}]}{3/5} = -\cos t\mathbf{j} - \sin t\mathbf{k}$$

Therefore

$$\mathbf{a}\left(\frac{\pi}{2}\right) = -3\mathbf{k},\ \mathbf{T}\left(\frac{\pi}{2}\right) = \frac{1}{5}[4\mathbf{i} - 3\mathbf{j}], \text{ and } \mathbf{N}\left(\frac{\pi}{2}\right) = -\mathbf{k}.$$

Thus,

$$a_{\mathbf{T}} = \mathbf{a}\left(\frac{\pi}{2}\right) \cdot \mathbf{T}\left(\frac{\pi}{2}\right) = 0 \quad \text{and} \quad a_{\mathbf{N}} = \mathbf{a}\left(\frac{\pi}{2}\right) \cdot \mathbf{N}\left(\frac{\pi}{2}\right) = 3.$$

**41.** $\mathbf{r}(t) = (v_0 t\cos\theta)\mathbf{i} + (h + v_0 t\sin\theta - 16t^2)\mathbf{j}$

$\mathbf{v}(t) = (v_0\cos\theta)\mathbf{i} + (v_0\sin\theta - 32t)\mathbf{j}$

$\mathbf{a}(t) = -32\mathbf{j}$

$$\mathbf{T}(t) = \frac{\mathbf{v}(t)}{\|\mathbf{v}(t)\|} = \frac{(v_0\cos\theta)\mathbf{i} + (v_0\sin\theta - 32t)\mathbf{j}}{\sqrt{v_0^2\cos^2\theta + (v_0\sin\theta - 32t)^2}}$$

Since the path of a projectile is concave downward,

$$\mathbf{N}(t) = \frac{(v_0\sin\theta - 32t)\mathbf{i} + (-v_0\cos\theta)\mathbf{j}}{\sqrt{v_0^2\cos^2\theta + (v_0\sin\theta - 32t)^2}}.$$

Therefore,

$$a_{\mathbf{T}} = \frac{-32(v_0\sin\theta - 32t)}{\sqrt{v_0^2\cos^2\theta + (v_0\sin\theta - 32t)^2}}$$

$$a_{\mathbf{N}} = \frac{32\,v_0\cos\theta}{\sqrt{v_0^2\cos^2\theta + (v_0\sin\theta - 32t)^2}}.$$

The projectile will reach its maximum height when the vertical component of velocity is zero, or

$v_0\sin\theta - 32t = 0.$

Hence, at the maximum height $a_{\mathbf{T}} = 0$ and $a_{\mathbf{N}} = 32$. Thus at the maximum height of the projectile, all the acceleration is normal to the path.

**43.** $\mathbf{r}(t) = a\cos(\omega t)\mathbf{i} + a\sin(\omega t)\mathbf{j}$

$\mathbf{v}(t) = -a\omega\sin(\omega t)\mathbf{i} + a\omega\cos(\omega t)\mathbf{j}$

$\mathbf{a}(t) = -a\omega^2\cos(\omega t)\mathbf{i} - a\omega^2\sin(\omega t)\mathbf{j}$

$$\mathbf{T}(t) = \frac{\mathbf{v}(t)}{\|\mathbf{v}(t)\|} = -\sin(\omega t)\mathbf{i} + \cos(\omega t)\mathbf{j}$$

$$\mathbf{N}(t) = \frac{\mathbf{T}'(t)}{\|\mathbf{T}'(t)\|} = -\cos(\omega t)\mathbf{i} - \sin(\omega t)\mathbf{j}$$

$a_{\mathbf{T}} = \mathbf{a} \cdot \mathbf{T} = 0$

$a_{\mathbf{N}} = \mathbf{a} \cdot \mathbf{N} = a\omega^2$

(a) If $\omega_0 = 2\omega$, then

$$\mathbf{a} \cdot \mathbf{N} = a\omega_0^2 = a(2\omega^2)^2 = 4aw^2.$$

Therefore, the centripetal acceleration is increased by a factor of 4 when the velocity is doubled.

(b) If $a_0 = a/2$, then

$$\mathbf{a} \cdot \mathbf{N} = a_0\omega^2 = \frac{a}{2}(\omega^2) = \frac{1}{2}aw^2.$$

Therefore, the centripetal acceleration is halved when the radius is halved.

## Section 11.5 Arc Length and Curvature

**7.** $\mathbf{r}(t) = a\cos t\mathbf{i} + a\sin t\mathbf{j} + bt\mathbf{k}$

The graph is a circular helix of radius $a$ (see figure). Since

$$\mathbf{r}'(t) = -a\sin t\mathbf{i} + a\cos t\mathbf{j} + b\mathbf{k},$$

the arc length on the interval $[0, 2\pi]$ is

$$s = \int_0^{2\pi} \|\mathbf{r}'(t)\|\,dt$$

$$= \int_0^{2\pi} \sqrt{(-a\sin t)^2 + (a\cos t)^2 + b^2}\,dt$$

$$= \int_0^{2\pi} \sqrt{a^2(\sin^2 t + \cos^2 t) + b^2}\,dt$$

$$= \int_0^{2\pi} \sqrt{a^2 + b^2}\,dt = \Big[\sqrt{a^2+b^2}\,t\Big]_0^{2\pi} = 2\pi\sqrt{a^2+b^2}.$$

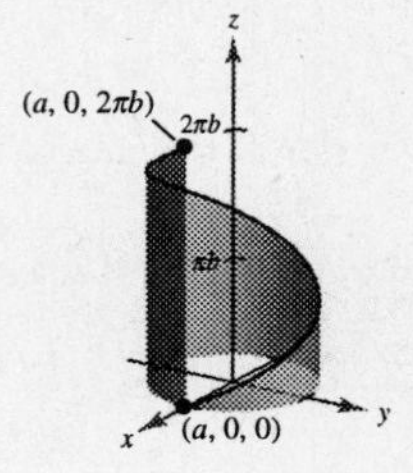


**13.** $\mathbf{r}(t) = \langle 2\cos t, 2\sin t, t\rangle$

(a) Since $x'(t) = -2\sin t$, $y'(t) = 2\cos t$, and $z'(t) = 1$, we have

$$s = \int_0^t \sqrt{[x'(\tau)]^2 + [y'(\tau)]^2 + [z'(\tau)]^2}\,d\tau = \int_0^t \sqrt{(-2\sin\tau)^2 + (2\cos\tau)^2 + (1)^2}\,d\tau$$

$$= \int_0^t \sqrt{5}\,d\tau = \sqrt{5}\Big[\tau\Big]_0^t = \sqrt{5}t.$$

(b) Since $s = \sqrt{5}t$, we have $t = s/\sqrt{5}$. Therefore, the parameterization of th curve in terms of $s$ is

$$r(s) = \left\langle 2\cos\frac{s}{\sqrt{5}}, 2\sin\frac{s}{\sqrt{5}}, \frac{s}{\sqrt{5}}\right\rangle.$$

(c) $\mathbf{r}(\sqrt{5}) = \langle 2\cos 1, 2\sin 1, 1\rangle \approx \langle 1.081, 1.683, 1\rangle$

and

$$\mathbf{r}(4) = \left\langle 2\cos\frac{4}{\sqrt{5}}, 2\sin\frac{4}{\sqrt{5}}, 1\right\rangle \approx \langle -0.433, 1.953, 1.789\rangle.$$

(d) $$\mathbf{r}'(s) = \left\langle -\frac{2}{\sqrt{5}}\sin\frac{s}{\sqrt{5}}, \frac{2}{\sqrt{5}}\cos\frac{s}{\sqrt{5}}, \frac{1}{\sqrt{5}}\right\rangle$$

$$\|\mathbf{r}'(s)\| = \sqrt{\left(-\frac{2}{\sqrt{5}}\sin\frac{s}{\sqrt{5}}\right)^2 + \left(\frac{2}{\sqrt{5}}\cos\frac{s}{\sqrt{5}}\right)^2 + \left(\frac{1}{\sqrt{5}}\right)^2} = \sqrt{\frac{4}{5}+\frac{1}{5}} = 1$$

**21.** $\mathbf{r}(t) = t\mathbf{i} + \dfrac{1}{t}\mathbf{j}$

From Exercise 17, Section 11.4, we have

$$\mathbf{a}(1)\cdot\mathbf{N}(1) = \sqrt{2} \quad\text{and}\quad \|\mathbf{v}(1)\|^2 = 2.$$

Therefore, the curvature is

$$K = \frac{\mathbf{a}(1)\cdot\mathbf{N}(1)}{\|\mathbf{v}(1)\|^2} = \frac{\sqrt{2}}{2} \approx 0.707.$$

**27.** $$\mathbf{r}(t) = e^t\cos t\mathbf{i} + e^t\sin t\mathbf{j}$$

$$\mathbf{r}'(t) = e^t(\cos t - \sin t)\mathbf{i} + e^t(\cos t + \sin t)\mathbf{j}$$

$$\|\mathbf{r}'(t)\| = e^t\sqrt{(\cos t - \sin t)^2 + (\cos t + \sin t)^2} = \sqrt{2}e^t$$

$$\mathbf{T}(t) = \frac{\mathbf{r}'(t)}{\|\mathbf{r}'(t)\|} = \frac{1}{\sqrt{2}}[(\cos t - \sin t)\mathbf{i} + (\cos t + \sin t)\mathbf{j}]$$

$$\mathbf{T}'(t) = \frac{1}{\sqrt{2}}[(-\sin t - \cos t)\mathbf{i} + (-\sin t + \cos t)\mathbf{j}]$$

$$\|\mathbf{T}'(t)\| = \frac{1}{\sqrt{2}}\sqrt{(-\sin t - \cos t)^2 + (-\sin t + \cos t)^2} = 1$$

$$K = \frac{\|\mathbf{T}'(t)\|}{\|\mathbf{r}'(t)\|} = \frac{1}{\sqrt{2}e^t} = \frac{\sqrt{2}}{2}e^{-t}$$

**33.** $\mathbf{r}(t) = 4t\mathbf{i} + 3\cos t\mathbf{j} + 3\sin t\mathbf{k}$

From Exercise 35, Section 11.4, we have

$$\|\mathbf{T}'(t)\| = \frac{3}{5} \quad \text{and} \quad \|\mathbf{r}'(t)\| = 5.$$

Therefore, the curvature is

$$K = \frac{\|\mathbf{T}'(t)\|}{\|\mathbf{r}'(t)\|} = \frac{3/5}{5} = \frac{3}{25}.$$

**37.** $y = 2x^2 + 3, y' = 4x, y'' = 4$

$$K = \left|\frac{y''}{[1 + (y')^2]^{3/2}}\right| = \frac{4}{[1 + (4x)^2]^{3/2}}$$

When $x = -1$, the curvature is

$$K = \frac{4}{(1 + 16)^{3/2}} = \frac{4}{17^{3/2}} \approx 0.057$$

and the radius of curvature when $x = -1$ is

$$r = \frac{1}{K} = \frac{17^{3/2}}{4} \approx 17.523.$$

**43.** $f(x) = x + \frac{1}{x}, f'(x) = 1 - \frac{1}{x^2} = \frac{x^2 - 1}{x^2}, f''(x) = \frac{2}{x^3}$

At the point $(1, 2), f'(1) = 0$, and $f''(1) = 2$. Thus at $(1, 2)$ the curvature is

$$K = \left|\frac{2}{(1 + 0^2)^{3/2}}\right| = 2$$

and the radius of curvature is

$$r = \frac{1}{K} = \frac{1}{2}.$$



Since the slope of the tangent line to the graph of the function is 0 at the point $(1, 2)$, the normal line is vertical and the center of the closest circular approximation is $\left(1, \frac{5}{2}\right)$ (see figure). Finally, the equation of the closest circular approximation is

$$(x - 1)^2 + \left(y - \frac{5}{2}\right)^2 = \left(\frac{1}{2}\right)^2.$$

**53.** $x^2 + 4y^2 = 4$

The endpoints of the major axis are $(\pm 2, 0)$ and the endpoints of the minor axis are $(0, \pm 1)$.

$$x^2 + 4y^2 = 4$$

$$2x + 8yy' = 0$$

$$y' = \frac{-x}{4y}$$

$$y'' = \frac{(4y)(-1) - (-x)(4y')}{16y^2} = \frac{-4y - (x^2/y)}{16y^2} = \frac{-(4y^2 + x^2)}{16y^3}$$

$$= \frac{-4}{16y^3} = \frac{-1}{4y^3}$$

The curvature is given by

$$K = \left|\frac{-1/4y^3}{[1 + (-x/4y)^2]^{3/2}}\right| = \left|\frac{-1}{4y^3[(16y^2 + x^2)/16y^2]^{3/2}}\right| = \left|\frac{-16}{(16y^2 + x^2)^{3/2}}\right|$$

$$= \frac{16}{(12y^2 + 4y^2 + x^2)^{3/2}} = \frac{16}{(12y^2 + 4)^{3/2}}.$$

Since $-1 \le y \le 1$, $K$ is greatest when $y = 0$ and smallest when $y = \pm 1$.

**63.** $r = 1 + \sin\theta$

$r' = \cos\theta$

$r'' = -\sin\theta$

$$K = \frac{|2(r')^2 - rr'' + r^2|}{[(r')^2 + r^2]^{3/2}}$$

$$= \frac{|2\cos^2\theta - (1+\sin\theta)(-\sin\theta) + (1+\sin\theta)^2|}{\sqrt{[\cos^2\theta + (1+\sin\theta)^2]^3}}$$

$$= \frac{3(1+\sin\theta)}{\sqrt{8(1+\sin\theta)^3}} = \frac{3}{2\sqrt{2(1+\sin\theta)}}$$

**73.** $x(\theta) = a(\theta - \sin\theta), y(\theta) = a(1 - \cos\theta)$

$x'(\theta) = a(1 - \cos\theta), y'(\theta) = a\sin\theta$

$x''(\theta) = a\sin\theta, y'' = a\cos\theta$

$$K = \frac{|x'(\theta)y''(\theta) - y'(\theta)x''(\theta)|}{\{[x'(\theta)]^2 + [y'(\theta)]^2\}^{3/2}}$$

$$= \frac{|a(1-\cos\theta)(a\cos\theta) - (a\sin\theta)(a\sin\theta)|}{\{[a(1-\cos\theta)]^2 + [a\sin\theta]^2\}^{3/2}}$$

$$= \frac{\sqrt{2}}{4a\sqrt{1-\cos\theta}} = \frac{1}{4a}\left|\csc\frac{\theta}{2}\right|$$

The minimum curvature is $K = 1/(4a)$ when $\theta = \pi$. There is no maximum since $K \to \infty$ as $\theta$ approaches even multiples of $\pi$.

## Review Exercises for Chapter 11

**7.** $\mathbf{r}(t) = \tan t\mathbf{i} - \sec t\mathbf{j} + (1 - t)\mathbf{k}$

$\mathbf{u}(t) = 2\cos t\mathbf{i} + 3\sin t\cos t\,\mathbf{j} + \mathbf{k}$

$$\mathbf{r}(t)\cdot\mathbf{u}(t) = (\tan t)(2\cos t) + (-\sec t)(3\sin t\cos t) + (1-t)(1)$$

$$= 2\sin t - 3\sin t + 1 - t = 1 - t - \sin t$$

The dot product of two vector-valued functions is a scalar function.

**17.** Label the sides of the triangle as shown in the figure. The line segment $C_1$ has slope $m = \frac{3}{4}$ and $y$-intercept $(0, 0)$. Hence, the rectangular form of the equation is

$$y = \tfrac{3}{4}x.$$

Letting $x = 4t$ and substituting this expression for $x$ in the rectangular equation of the line yields $y = 3t$. Therefore,

$$\mathbf{r}_1(t) = 4t\mathbf{i} + 3t\mathbf{j},\ 0 \le t \le 1.$$

Note that $x = 4$ on $C_2$. Since $y$ is decreasing on $C_2$, a vector-valued function for $C_2$ is

$$\mathbf{r}_2(t) = 4\mathbf{i} - (3 - t)\mathbf{j},\ 0 \le t \le 3.$$

Observe that $y = 0$ on $C_3$. Since $x$ is decreasing on $C_3$, a vector-valued function for $C_3$ is

$$\mathbf{r}_3(t) = (4 - t)\mathbf{i},\ 0 \le t \le 4.$$

(Observe that there are many correct answers for this problem.)

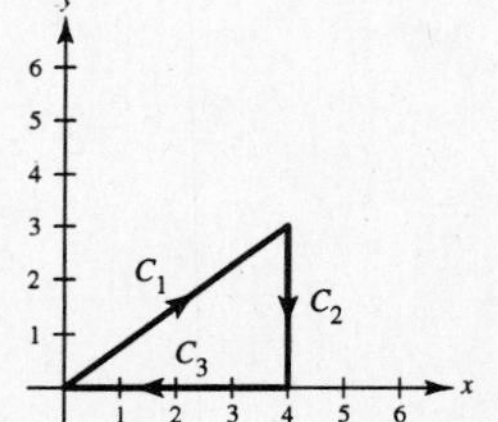


**27.** $\mathbf{r}(t) = 2\cos t\mathbf{i} + 2\sin t\mathbf{j} + t\mathbf{k}$

$$\mathbf{r}\left(\frac{3\pi}{4}\right) = -\sqrt{2}\mathbf{i} + \sqrt{2}\mathbf{j} + \frac{3\pi}{4}\mathbf{k}$$

$\mathbf{r}'(t) = -2\sin t\mathbf{i} + 2\cos t\mathbf{j} + \mathbf{k}$

$$\mathbf{r}'\left(\frac{3\pi}{4}\right) = -\sqrt{2}\mathbf{i} - \sqrt{2}\mathbf{j} + \mathbf{k}$$

Therefore, the tangent line must pass through the point $\left(-\sqrt{2}, \sqrt{2}, 3\pi/4\right)$ and have direction numbers $a = -\sqrt{2}$, $b = -\sqrt{2}$, and $c = 1$. Thus, the parametric equations of the line are given by

$$x = -\sqrt{2} - \sqrt{2}t,\ y = \sqrt{2} - \sqrt{2}t, \text{ and } z = \frac{3\pi}{4} + t.$$

**35.** $\int \|\cos t\mathbf{i} + \sin t\mathbf{j} + t\mathbf{k}\|\, dt = \int \sqrt{\cos^2 t + \sin^2 t + t^2}\, dt$

$$= \int \sqrt{1 + t^2}\, dt = \frac{1}{2}\left(t\sqrt{1 + t^2} + \ln\left|t + \sqrt{1 + t^2}\right|\right) + C$$

**43.** $\mathbf{r}(t) = (v_0 \cos\theta)t\mathbf{i} + [(v_0 \sin\theta)t - 16t^2]\mathbf{j}$

The projectile will strike the ground when

$$(v_0 \sin\theta)t - 16t^2 = 0$$

$$t(v_0 \sin\theta - 16t) = 0 \implies t = \frac{v_0 \sin\theta}{16}.$$

Substituting this expression for $t$ into the $x$-component of the vector-valued function will give the range.

Range: $x = v_0 \cos\theta\left(\frac{v_0 \sin\theta}{16}\right) = \left(\frac{v_0^2}{32}\right)\sin 2\theta$

When $\theta = 30°$ and $v_0 = 75$, the range is

$$x = \left(\frac{75^2}{32}\right)\sin 60° \approx 152 \text{ feet.}$$

**53.** $\mathbf{r}(t) = t\mathbf{i} + t^2\mathbf{j} + \frac{1}{2}t^2\mathbf{k}$

$$\mathbf{v}(t) = \mathbf{r}'(5) = \mathbf{i} + 2t\mathbf{j} + t\mathbf{k}$$

$$\text{speed} = \|\mathbf{v}(t)\| = \sqrt{1^2 + (2t)^2 + t^2} = \sqrt{1 + 5t^2}$$

$$\mathbf{a}(t) = \mathbf{r}''(t) = 2\mathbf{j} + \mathbf{k}$$

$$\mathbf{T}(t) = \frac{\mathbf{v}(t)}{\|\mathbf{v}(t)\|} = \frac{\mathbf{i} + 2t\mathbf{j} + t\mathbf{k}}{\sqrt{1 + 5t^2}}$$

$$\mathbf{T}'(t) = \frac{5t}{(1 + 5t^2)^{3/2}}\mathbf{i} + \frac{2}{(1 + 5t^2)^{3/2}}\mathbf{j} + \frac{1}{(1 + 5t^2)^{3/2}}\mathbf{k} = \frac{5t\mathbf{i} + 2\mathbf{j} + \mathbf{k}}{(1 + 5t^2)^{3/2}}$$

$$\|\mathbf{T}'(t)\| = \frac{\sqrt{(5t)^2 + 2^2 + 1^2}}{(1 + 5t^2)^{3/2}} = \frac{\sqrt{5}\sqrt{1 + 5t^2}}{(1 + 5t^2)^{3/2}} = \frac{\sqrt{5}}{1 + 5t^2}$$

$$\mathbf{N}(t) = \frac{\mathbf{T}'(t)}{\|\mathbf{T}'(t)\|} = \frac{5t\mathbf{i} + 2\mathbf{j} + \mathbf{k}}{\sqrt{5}\sqrt{1 + 5t^2}}$$

$$a_{\mathbf{T}} = \mathbf{a}(t) \cdot \mathbf{T}(t) = \frac{2(2t) + 1(t)}{\sqrt{1 + 5t^2}} = \frac{5t}{\sqrt{1 + 5t^2}}$$

$$a_{\mathbf{N}} = \mathbf{a}(t) \cdot \mathbf{N}(t) = \frac{2(2) + 1(1)}{\sqrt{5}\sqrt{1 + 5t^2}} = \frac{5}{\sqrt{5}\sqrt{1 + 5t^2}} = \frac{\sqrt{5}}{\sqrt{1 + 5t^2}}$$

$$K = \frac{\mathbf{a}(t) \cdot \mathbf{N}(t)}{\|\mathbf{v}(t)\|} = \frac{\frac{\sqrt{5}}{\sqrt{1 + 5t^2}}}{1 + 5t^2} = \frac{\sqrt{5}}{(1 + 5t^2)^{3/2}}$$

**55.** $\mathbf{r}(t) = \frac{1}{2}t\mathbf{i} + \sin t\mathbf{j} + \cos t\mathbf{k}$

$$s = \int_0^{\pi} \|\mathbf{r}'(t)\|\, dt$$

$$= \int_0^{\pi} \sqrt{(1/2)^2 + \cos^2 t + (-\sin t)^2} dt$$

$$= \frac{\sqrt{5}}{2}\int_0^{\pi} dt = \frac{\sqrt{5}}{2}\Big[t\Big]_0^{\pi} = \frac{\sqrt{5}\pi}{2}$$

# CHAPTER 12
# Functions of Several Variables

# CHAPTER 12
# Functions of Several Variables

## Section 12.1 Introduction to Functions of Several Variables

### Solutions to Selected Odd-Numbered Exercises

**7.** $f(x, y) = xe^y$

(a) $f(5, 0) = 5e^0 = 5$ (b) $f(3, 2) = 3e^2$

(c) $f(2, -1) = 2e^{-1} = \dfrac{2}{e}$ (d) $f(5, y) = 5e^y$

(e) $f(x, 2) = xe^2$ (f) $f(t, t) = te^t$

**13.** $f(x, y) = \displaystyle\int_x^y (2t - 3)\, dt = \Big[t^2 - 3t\Big]_x^y$

$= (y^2 - 3y) - (x^2 - 3x)$

(a) $f(0, 4) = (16 - 12) - (0 - 0) = 4$

(b) $f(1, 4) = (16 - 12) - (1 - 3) = 6$

**17.** Since $f(x, y) = \sqrt{4 - x^2 - y^2}$, we have

$$4 - x^2 - y^2 \geq 0$$
$$4 \geq x^2 + y^2.$$

Therefore, the domain is the set of all points inside and on the boundary of the circle $x^2 + y^2 = 4$. The range of $f$ is the set of all real numbers in the interval $[0, 2]$.

**19.** Since $z = \arcsin(x + y)$ implies that $\sin z = x + y$, we conclude that $|x + y| \leq 1$. Therefore, the domain is

$$-1 \leq x + y \leq 1$$
$$-1 - x \leq y \leq -x + 1.$$

This means that $R$ lies on and between the parallel lines

$$y = -1 - x \quad \text{and} \quad y = -x + 1$$

as shown in the figure. The range of the arcsine function is the set of all reals in the interval $[-\pi/2, \pi/2]$.

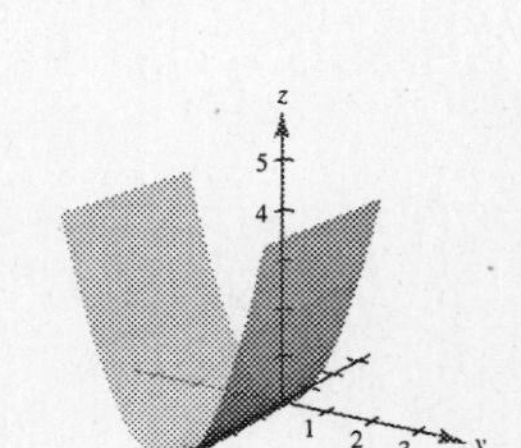


**33.** $f(x, y) = y^2$

Since the variable $x$ is missing, the surface is a cylinder with rulings parallel to the $x$-axis. The generating curve is $z = y^2$. The domain is the entire $xy$-plane and the range is $z \geq 0$.

**43.** $f(x, y) = x^2 + y^2$

(a) The surface is a paraboloid and its axis is the $z$-axis. Some conventional traces are

| | | |
|---|---|---|
| $yz$-trace $(x = 0)$: | $z = y^2$ | Parabola |
| $xz$-trace $(y = 0)$: | $z = x^2$ | Parabola |
| Parallel to $xy$-plane $(z = 4)$: | $4 = x^2 + y^2$ | Circle. |

The domain is the entire $xy$-coordinate plane and the range $z \geq 0$. The surface is shown in the figure.

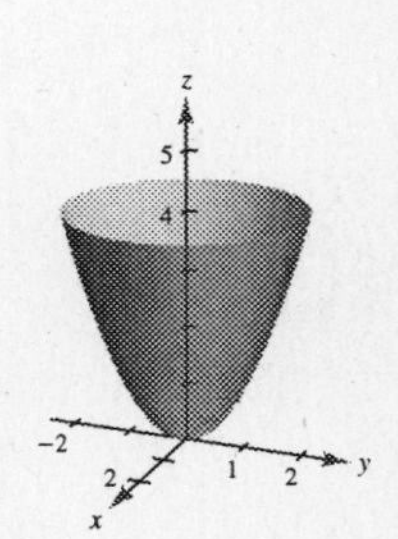


—CONTINUED—

**43. —CONTINUED—**

(b) The graph of $g$ is a vertical translation of the graph of $f$ two units upward.

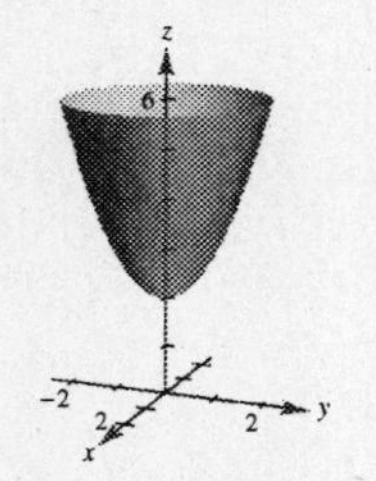

(c) The graph of $g$ is a horizontal translation of the graph of $f$ two units to the right.

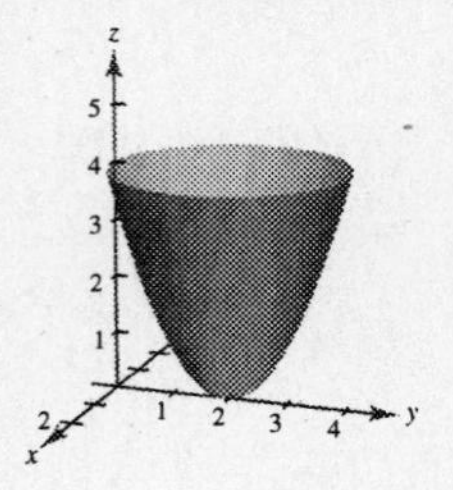

(d) The graph of $g$ is a reflection of the graph of $f$ in the $xy$-plane followed by a vertical translation four units upward.

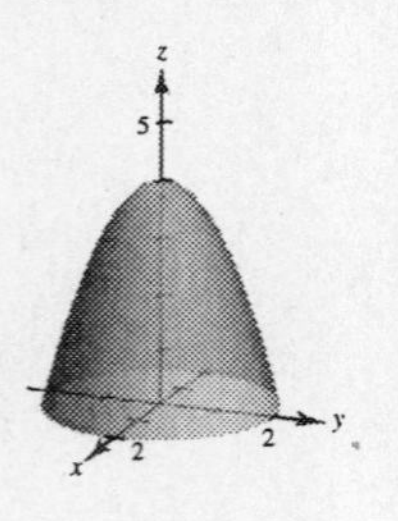

(e) The trace parallel to the $yz$-coordinate plane when $x = 1$ is the parabola $z = f(1, y) = 1 + y^2$. The trace parallel to the $xz$-coordinate when $y = 1$ is the parabola

$$z = f(x, 1) = x^2 + 1.$$

The traces are shown in the figures.

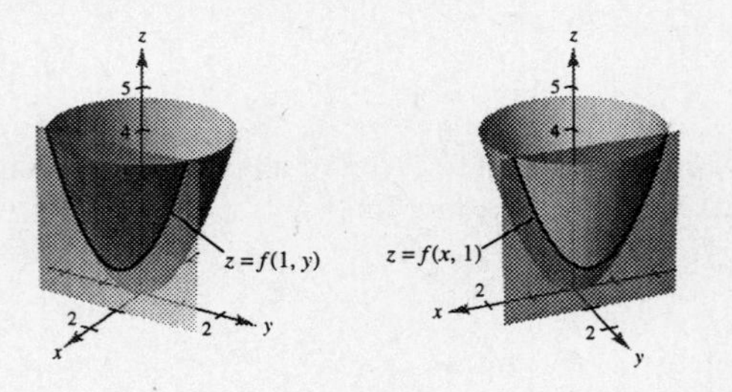


**55.** $f(x, y) = \dfrac{x}{x^2 + y^2}$

If $f(x, y) = c$, then the level curves are of the form

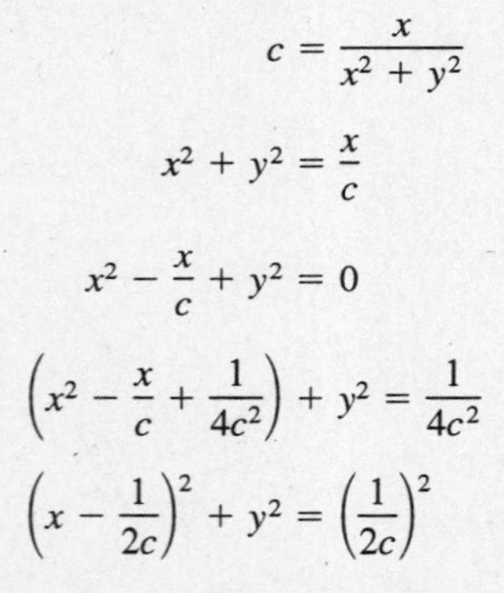

$$c = \frac{x}{x^2 + y^2}$$

$$x^2 + y^2 = \frac{x}{c}$$

$$x^2 - \frac{x}{c} + y^2 = 0$$

$$\left(x^2 - \frac{x}{c} + \frac{1}{4c^2}\right) + y^2 = \frac{1}{4c^2}$$

$$\left(x - \frac{1}{2c}\right)^2 + y^2 = \left(\frac{1}{2c}\right)^2$$

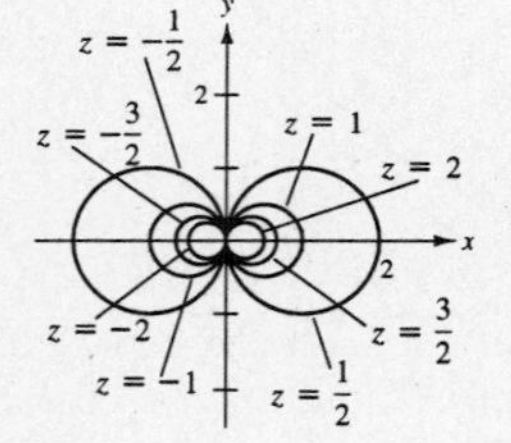


Therefore, each level curve is a circle centered at $(1/(2c), 0)$ with radius equal to $1/(2c)$. For example, if $c = 1$, the level curve has the equation

$$\left(x - \frac{1}{2}\right)^2 + y^2 = \frac{1}{4}.$$

The required level curves are shown in the figure.

**67.** $f(x, y, z) = 4x^2 + 4y^2 - z^2$

The level surface when $f(x, y, z) = 0$ is the cone

$$4x^2 + 4y^2 - z^2 = 0$$

$$x^2 = 4x^2 + 4y^2.$$

$xz$-trace: $z = \pm 2x$

$yz$-trace: $z = \pm 2y$

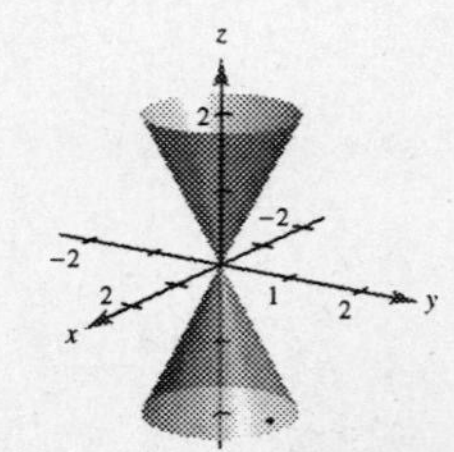

**73.** Assign variables to the length, width, and height of the box as shown in the figure.

$$C = (\text{cost of base}) + (\text{cost of front and back}) + (\text{cost of two ends})$$
$$= 0.75xy + 2(0.40)xz + 2(0.40)yz$$
$$= 0.75xy + 0.80(xz + yz)$$

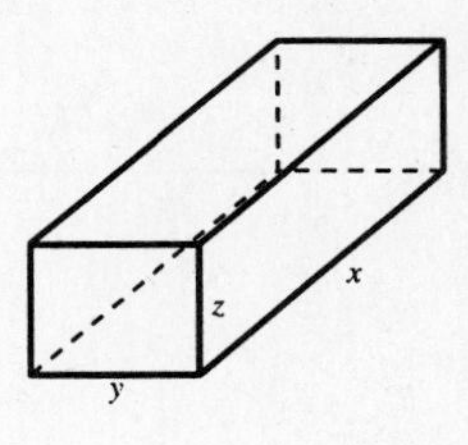


## Section 12.2 Limits and Continuity

**9.** $\displaystyle\lim_{(x,y)\to(0,1)} \frac{\arcsin(x/y)}{1+xy}$

Since the limit of a quotient is the quotient of the limits, we have

$$\lim_{(x,y)\to(0,1)} \frac{\arcsin(x/y)}{1+xy} = \frac{\arcsin 0}{1+0} = \frac{0}{1} = 0.$$

A rational function is continuous at every point in its domain. Therefore, the given function is continuous for all points $(x, y)$ in the $xy$-plane such that $1 + xy \neq 0$, $y \neq 0$, and $|x/y| \leq 1$.

**21.** $f(x, y) = -\dfrac{xy^2}{x^2+y^2}$

Path: $x = y^2$

| $x$ | $(1, 1)$ | $(0.25, 0.5)$ | $(0.01, 0.1)$ | $(0.0001, 0.011)$ | $(0.000001, 0.0011)$ |
|---|---|---|---|---|---|
| $f(x, y)$ | $-\frac{1}{2}$ | $-\frac{1}{2}$ | $-\frac{1}{2}$ | $-\frac{1}{2}$ | $-\frac{1}{2}$ |

$$\lim_{(x,y)\to(0,0)} \frac{-xy^2}{x^2+y^4} = \lim_{(y^2,y)\to(0,0)} \frac{-y^2(y^2)}{(y^2)^2+y^4}$$
$$= \lim_{(y^2,y)\to(0,0)} \frac{-y^4}{y^4+y^4} = -\frac{1}{2}$$

Path: $x = -y^2$

| $x$ | $(1, 1)$ | $(-0.25, 0.5)$ | $(-0.01, 0.1)$ | $(-0.0001, 0.011)$ | $(-0.000001, 0.0011)$ |
|---|---|---|---|---|---|
| $f(x, y)$ | $\frac{1}{2}$ | $\frac{1}{2}$ | $\frac{1}{2}$ | $\frac{1}{2}$ | $\frac{1}{2}$ |

$$\lim_{(x,y)\to(0,0)} \frac{-xy^2}{x^2+y^4} = \lim_{(-y^2,y)\to(0,0)} \frac{-(-y^2)(y^2)}{(-y^2)^2+y^4}$$
$$= \lim_{(-y^2,y)\to(0,0)} \frac{y^4}{y^4+y^4} = \frac{1}{2}.$$

Since the limits are not the same along different paths, the limit does not exist. The function is continuous except at $(0, 0)$.

**29.** $\displaystyle\lim_{(x,y)\to(0,0)} \frac{\sin(x^2+y^2)}{x^2+y^2}$

We first observe that direct substitution yields the indeterminate form $0/0$. Letting $x = r\cos\theta$, $y = r\sin\theta$, and $r^2 = x^2 + y^2$, yields

$$\lim_{(x,y)\to(0,0)} \frac{\sin(x^2+y^2)}{x^2+y^2} = \lim_{r\to 0} \frac{\sin r^2}{r^2} = 1.$$

**39.** $f(t) = \dfrac{1}{t}$, $g(x, y) = 3x - 2y$

$$(f \circ g)(x, y) = f[g(x, y)]$$
$$= \frac{1}{g(x, y)} = \frac{1}{3x - 2y}$$

The composite function is continuous for $y \neq 3x/2$.

**41.** $f(x, y) = x^2 - 4y$

(a) $$\lim_{\Delta x \to 0} \frac{f(x + \Delta x, y) - f(x, y)}{\Delta x} = \lim_{\Delta x \to 0} \frac{[(x + \Delta x)^2 - 4y] - (x^2 - 4y)}{\Delta x}$$

$$= \lim_{\Delta x \to 0} \frac{x^2 + 2x\Delta x + (\Delta x)^2 - 4y - x^2 + 4y}{\Delta x}$$

$$= \lim_{\Delta x \to 0} (2x + \Delta x) = 2x$$

(b) $$\lim_{\Delta y \to 0} \frac{f(x, y + \Delta y) - f(x, y)}{\Delta y} = \lim_{\Delta y \to 0} \frac{x^2 - 4(y + \Delta y) - (x^2 - 4y)}{\Delta y}$$

$$= \lim_{\Delta y \to 0} \frac{x^2 - 4y - 4\Delta y - x^2 + 4y}{\Delta y}$$

$$= \lim_{\Delta y \to 0} (-4) = -4$$

## Section 12.3 Partial Derivatives

**7.** $z = x\sqrt{y} = xy^{1/2}$

Considering $y$ to be a constant and differentiating with respect to $x$ yields

$$\frac{\partial z}{\partial x} = (1)y^{1/2} = \sqrt{y}.$$

Considering $x$ to be a constant and differentiating with respect to $y$ yields

$$\frac{\partial z}{\partial y} = x\left(\frac{1}{2}\right)y^{-1/2} = \frac{x}{2\sqrt{y}}.$$

**13.** $z = \ln \dfrac{x + y}{x - y}$

Using the properties of the logarithm function, rewrite the function to obtain

$$z = \ln(x + y) - \ln(x - y).$$

Considering $y$ to be a constant and differentiating with respect to $x$ we have

$$\frac{\partial z}{\partial x} = \frac{1}{x + y}(1) - \frac{1}{x - y}(1) = \frac{-2y}{x^2 - y^2}.$$

Now considering $x$ to be a constant and differentiating with respect to $y$ we have

$$\frac{\partial z}{\partial y} = \frac{1}{x + y}(1) - \frac{1}{x - y}(-1) = \frac{2x}{x^2 - y^2}.$$

**21.** $z = e^y \sin xy$

First, considering $y$ to be constant, we have

$$\frac{\partial z}{\partial x} = e^y(\cos xy)(y) = ye^y \cos xy.$$

Now considering $x$ to be constant and using the Product Rule, we have

$$\frac{\partial z}{\partial y} = e^y(\cos xy)(x) + (\sin xy)(e^y)(1) = e^y(x \cos xy + \sin xy).$$

**29.** $g(x, y) = 4 - x^2 - y^2$

Considering $y$ as a constant and differentiating with respect to $x$ produces

$$g_x(x, y) = -2x.$$

Differentiating with respect to $y$ and considering $x$ as a constant yields

$$g_y(x, y) = -2y.$$

The slope of the surface in the $x$ direction at $(1, 1, 2)$ is $g_x(1, 1) = -2$. The slope of the surface in the $y$ direction at $(1, 1, 2)$ is $g_y(1, 1) = -2$.

**33.** $f(x, y) = \arctan \frac{y}{x}$

First, considering $y$ to be constant, we have

$$\frac{\partial z}{\partial x} = \frac{1}{1 + (y^2/x^2)}\left(\frac{-y}{x^2}\right) = \frac{-y}{x^2 + y^2}.$$

Now considering $x$ to be constant, we have

$$\frac{\partial z}{\partial y} = \frac{1}{1 + (y^2/x^2)}\left(\frac{1}{x}\right) = \frac{x}{x^2 + y^2}.$$

Therefore, $f_x(2, -2) = \frac{1}{4}$ and $f_y(2, -2) = \frac{1}{4}$.

**41.** The graph of the equation $z = 9x^2 - y^2$ is a hyperbolic paraboloid. The $xy$-trace ($z = 0$), consists of the intersecting lines $y = \pm 3x$. The $yz$-trace ($x = 0$) is a parabola $z = -y^2$ opening downward, and the $xz$-trace ($y = 0$) is the parabola $z = 9x^2$ opening upward. The curve of intersection of the paraboloid and plane $y = 3$ is given by $z = 9x^2 - 9$. It is a parabola opening upward (see figure). Since $y$ is a constant on the curve of intersection, differentiate with respect to $x$ to obtain

$$\frac{\partial z}{\partial x} = 18x.$$

At the point $(1, 3, 0)$ the slope is

$$\frac{\partial z}{\partial x} = 18(1) = 18.$$

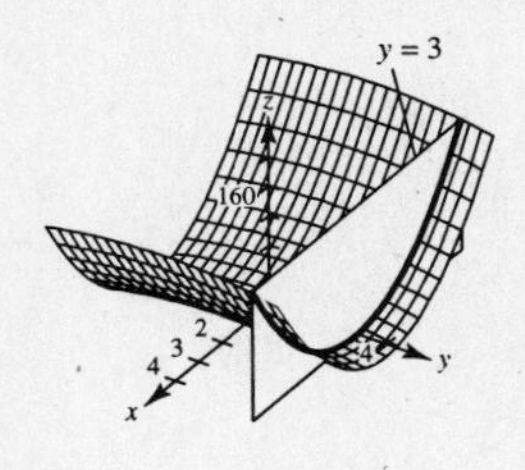


**47.** $z = x^2 - 2xy + 3y^2$

The first partials are

$$\frac{\partial z}{\partial x} = 2x - 2y \quad \text{and} \quad \frac{\partial z}{\partial y} = -2x + 6y.$$

The second partials are

$$\frac{\partial^2 z}{\partial x^2} = \frac{\partial}{\partial x}\left[\frac{\partial z}{\partial x}\right] = 2 \qquad \frac{\partial^2 z}{\partial y \partial x} = \frac{\partial}{\partial y}\left[\frac{\partial z}{\partial x}\right] = -2$$

$$\frac{\partial^2 z}{\partial y^2} = \frac{\partial}{\partial y}\left[\frac{\partial z}{\partial y}\right] = 6 \qquad \frac{\partial^2 z}{\partial x \partial y} = \frac{\partial}{\partial x}\left[\frac{\partial z}{\partial y}\right] = -2.$$

(Note that the mixed partial derivatives are equal.)

**53.** $z = \arctan \frac{y}{x}$

The first partial derivatives were found in Exercise 33.

$$\frac{\partial z}{\partial x} = \frac{-y}{x^2 + y^2} \quad \text{and} \quad \frac{\partial z}{\partial y} = \frac{x}{x^2 + y^2}$$

The second partials are

$$\frac{\partial^2 z}{\partial x^2} = \frac{\partial}{\partial x}\left[\frac{\partial z}{\partial x}\right] = \frac{(x^2 + y^2)(0) - (-y)(2x)}{(x^2 + y^2)^2} = \frac{2xy}{(x^2 + y^2)^2}$$

$$\frac{\partial^2 z}{\partial y \partial x} = \frac{\partial}{\partial y}\left[\frac{\partial z}{\partial x}\right] = \frac{(x^2 + y^2)(-1) - (-y)(2y)}{(x^2 + y^2)^2} = \frac{y^2 - x^2}{(x^2 + y^2)^2}$$

$$\frac{\partial^2 z}{\partial y^2} = \frac{\partial}{\partial y}\left[\frac{\partial z}{\partial y}\right] = \frac{(x^2 + y^2)(0) - x(2y)}{(x^2 + y^2)^2} = \frac{-2xy}{(x^2 + y^2)^2}$$

$$\frac{\partial^2 z}{\partial x \partial y} = \frac{\partial}{\partial x}\left[\frac{\partial z}{\partial y}\right] = \frac{(x^2 + y^2)(1) - x(2x)}{(x^2 + y^2)^2} = \frac{y^2 - x^2}{(x^2 + y^2)^2}.$$

(Note that the mixed partial derivatives are equal.)

**59.** $w = \sqrt{x^2 + y^2 + z^2} = (x^2 + y^2 + z^2)^{1/2}$

$$\frac{\partial w}{\partial x} = \frac{1}{2}(x^2 + y^2 + z^2)^{-1/2}(2x) = \frac{x}{\sqrt{x^2 + y^2 + z^2}}$$

$$\frac{\partial w}{\partial y} = \frac{1}{2}(x^2 + y^2 + z^2)^{-1/2}(2y) = \frac{y}{\sqrt{x^2 + y^2 + z^2}}$$

$$\frac{\partial w}{\partial z} = \frac{1}{2}(x^2 + y^2 + z^2)^{-1/2}(2z) = \frac{z}{\sqrt{x^2 + y^2 + z^2}}$$

**67.** $f(x, y, z) = e^{-x} \sin yz$

$$f_x(x, y, z) = -e^{-x} \sin yz$$

$$f_{xy}(x, y, z) = -e^{-x}(\cos yz)(z) = -ze^{-x} \cos yz$$

$$f_{xyy}(x, y, z) = -ze^{-x}(-\sin yz)(z) = z^2e^{-x} \sin yz$$

$$f_y(x, y, z) = e^{-x}(\cos yz)(z) = ze^{-x} \cos yz$$

$$f_{yx}(x, y, z) = -ze^{-x} \cos yz$$

$$f_{yxy}(x, y, z) = -ze^{-x}(-\sin yz)(z) = z^2e^{-x} \sin yz$$

$$f_y(x, y, z) = e^{-x}(\cos yz)(z) = ze^{-x} \cos yz$$

$$f_{yy}(x, y, z) = ze^{-x}(-\sin yz)(z) = -z^2e^{-x} \sin yz$$

$$f_{yyx}(x, y, z) = -z^2e^{-x}(-1) \sin yz = z^2e^{-x} \sin yx$$

Therefore,

$$f_{xyy}(x, y, z) = f_{yxy}(x, y, z) = f_{yyx}(x, y, z) = z^2e^{-x} \sin yx.$$

**73.** $z = \sin(x - ct)$

$$\frac{\partial z}{\partial x} = \cos(x - ct) \quad \text{and} \quad \frac{\partial^2 z}{\partial x^2} = -\sin(x - ct)$$

$$\frac{\partial z}{\partial t} = -c \cos(x - ct) \quad \text{and} \quad \frac{\partial^2 z}{\partial t^2} = -c^2 \sin(x - ct)$$

Therefore,

$$\frac{\partial^2 z}{\partial t^2} = -c^2 \sin(x - ct) = c^2 \frac{\partial^2 z}{\partial x^2}.$$

**79.** Let $N$ be the number of applicants to a university, $p$ the charge for food and housing, and $t$ the tuition. Since

$$\frac{\partial N}{\partial p} < 0 \quad \text{and} \quad \frac{\partial N}{\partial t} < 0,$$

it follows that an increase in either price will cause a decrease in the number of applicants.

## Section 12.4 Differentials

**5.** $z = x \cos y - y \cos x$

$$dz = \frac{\partial z}{\partial x}\,dx + \frac{\partial z}{\partial y}\,dy$$

$$= (\cos y + y \sin x)\,dx + (-x \sin y - \cos x)\,dy$$

$$= (\cos y + y \sin x)\,dx - (x \sin y + \cos x)\,dy$$

**9.** Since $u = (x + y)/(z - 2y)$, we have

$$du = \frac{\partial u}{\partial x}\,dx + \frac{\partial u}{\partial y}\,dy + \frac{\partial u}{\partial z}\,dz$$

$$= \frac{1}{z - 2y}\,dx + \frac{(z - 2y)(1) - (x + y)(-2)}{(z - 2y)^2}\,dy + \frac{0 - (x + y)(1)}{(z - 2y)^2}\,dz$$

$$= \frac{1}{z - 2y}\,dx + \frac{2x + z}{(z - 2y)^2}\,dy - \frac{x + y}{(z - 2y)^2}\,dz.$$

**11.** $f(x, y) = 9 - x^2 - y^2$

(a) $f(1, 2) = 9 - 1^2 - 2^2 = 4$

$f(1.05, 2.1) = 9 - (1.05)^2 - (2.1)^2 = 3.4875$

$\Delta z = f(1.05, 2.1) - f(1, 2) = -0.5125$

(b) $dz = f_x(x, y)\,dx + f_y(x, y)\,dy$

$= -2x\,dx - 2y\,dy$

Letting $x = 1$, $y = 2$, $dx = 0.05$, and $dy = 0.1$, yields

$dz = -2(1)(0.05) - 2(2)(0.1) = -0.5.$

**17.** Let $z = \sqrt{x^2 + y^2}$, $x = 5$, $y = 3$, $dx = 0.05$, and $dy = 0.1$. Then,

$$dz = \frac{\partial z}{\partial x}\,dx + \frac{\partial z}{\partial y}\,dy$$

$$= \frac{x}{\sqrt{x^2 + y^2}}\,dx + \frac{y}{\sqrt{x^2 + y^2}}\,dy$$

$$\sqrt{(5.05)^2 + (3.1)^2} - \sqrt{5^2 + 3^2} \approx \frac{5}{\sqrt{5^2 + 3^2}}(0.05) + \frac{3}{\sqrt{5^2 + 3^2}}(0.1) = \frac{0.55}{\sqrt{34}} \approx 0.094.$$

**25.** First consider the relative errors in $r$ and $h$ as

$$\frac{dr}{r} = \pm 4\% = \pm 0.04 \quad \text{and} \quad \frac{dh}{h} = \pm 2\% = \pm 0.02.$$

Since $V = \pi r^2 h$, we have

$$dV = 2\pi rh\,dr + \pi r^2\,dh$$

or the relative error in $V$ is

$$\frac{dV}{V} = \frac{2\pi rh\,dr}{\pi r^2 h} + \frac{\pi r^2\,dh}{\pi r^2 h} = 2\frac{dr}{r} + \frac{dh}{h}$$

$$= 2(\pm 0.04) \pm 0.02 = \pm 0.10 = \pm 10\%.$$

**33.** $L = 0.00021\left(\ln \frac{2h}{r} - 0.75\right)$

Using the properties of logarithms, rewrite the function as

$$L = 0.00021[\ln(2h) - \ln r - 0.75].$$

Letting $h = 100$, $r = 2$, $dh = \pm\frac{1}{100}$, and $dr = \pm\frac{1}{16}$, the total differential is given by

$$dL = 0.00021\left(\frac{1}{h}\,dh - \frac{1}{r}\,dr\right)$$

$$= 0.00021\left[\frac{1}{100}\left(\pm\frac{1}{100}\right) - \frac{1}{2}\left(\pm\frac{1}{16}\right)\right]$$

$$\approx \pm 6.6 \times 10^{-6} \text{ microhenrys.}$$

Approximating the inductance, we have

$$L = 0.00021(\ln 200 - \ln 2 - 0.75) \pm dL$$

$$= 8.096 \times 10^{-4} \pm 6.6 \times 10^{-6}.$$

**39.** $f(x, y) = x^2 - 2x + y$

$$\Delta z = f(x + \Delta x, y + \Delta y) - f(x, y)$$

$$= (x + \Delta x)^2 - 2(x + \Delta x) + (y + \Delta y) - (x^2 - 2x + y)$$

$$= x^2 + 2x(\Delta x) + (\Delta x)^2 - 2x - 2\Delta x + y + \Delta y - x^2 + 2x - y$$

$$= (2x - 2)\Delta x + (1)\Delta y + \Delta x(\Delta x) + 0(\Delta y)$$

$$= f_x(x, y)\Delta x + f_y(x, y)\Delta y + \epsilon_1(\Delta x) + \epsilon_2(\Delta y)$$

Therefore, $\epsilon_1 = \Delta x$ and $\epsilon_2 = 0$. As $(\Delta x, \Delta y) \to (0, 0)$, $\epsilon_1 \to 0$ and $\epsilon_2 \to 0$.

## Section 12.5 Chain Rules for Functions of Several Variables

**9.** $w = xy + xz + yz, x = t - 1, y = t^2 - 1, z = t$

(a) By the Chain Rule we have

$$\begin{aligned}\frac{dw}{dt} &= \frac{\partial w}{\partial x}\frac{dx}{dt} + \frac{\partial w}{\partial y}\frac{dy}{dt} + \frac{\partial w}{\partial z}\frac{dz}{dt}\\ &= (y + z)(1) + (x + z)(2t) + (x + y)(1)\\ &= (t^2 - 1 + t)(1) + (t - 1 + t)(2t) + (t - 1 + t^2 - 1)(1)\\ &= 6t^2 - 3 = 3(2t^2 - 1).\end{aligned}$$

(b) By writing $w$ as a function of $t$ before differentiating, we have

$$w = (t - 1)(t^2 - 1) + (t - 1)t + (t^2 - 1)t = 2t^3 - 3t + 1$$

$$\frac{dw}{dt} = 6t^2 - 3 = 3(2t^2 - 1).$$

**15.** $w = x^2 - y^2, x = s \cos t, y = s \sin t$

By the Chain Rule,

$$\frac{\partial w}{\partial s} = \frac{\partial w}{\partial x}\frac{\partial x}{\partial s} + \frac{\partial w}{\partial y}\frac{\partial y}{\partial s} \quad \text{and} \quad \frac{\partial w}{\partial t} = \frac{\partial w}{\partial x}\frac{\partial x}{\partial t} + \frac{\partial w}{\partial y}\frac{\partial y}{\partial t}.$$

Therefore,

$$\begin{aligned}\frac{\partial w}{\partial s} &= 2x(\cos t) + (-2y)(\sin t)\\ &= (2s \cos t)(\cos t) - (2s \sin t)(\sin t)\\ &= 2s(\cos^2 t - \sin^2 t) = 2s \cos 2t.\end{aligned}$$

When $s = 3$ and $t = \pi/4$, $\partial w/\partial s = 2(3)\cos(\pi/2) = 0$. Similarly,

$$\begin{aligned}\frac{\partial w}{\partial t} &= 2x(-s \sin t) - 2y(s \cos t) = 2s \cos t(-s \sin t) - 2s \sin t(s \cos t)\\ &= -4s^2 \sin t \cos t = -2s^2 \sin 2t.\end{aligned}$$

When $s = 3$ and $t = \pi/4$, $\partial w/\partial t = -2(9)\sin(\pi/2) = -18$.

**19.** $w = \arctan \dfrac{y}{x}, x = r \cos \theta, y = r \sin \theta$

(a) First calculate $\partial w/\partial x$ and $\partial w/\partial y$.

$$w = \arctan \frac{y}{x}$$

$$\frac{\partial w}{\partial x} = \frac{-y/x^2}{1 + (y^2/x^2)} = \frac{-y/x^2}{(x^2 + y^2)/x^2} = \frac{-y}{x^2 + y^2}$$

$$\frac{\partial w}{\partial y} = \frac{1/x}{1 + (y^2/x^2)} = \frac{1/x}{(x^2 + y^2)/x^2} = \frac{x}{x^2 + y^2}$$

Using the Chain Rule yields,

$$\begin{aligned}\frac{\partial w}{\partial r} &= \frac{\partial w}{\partial x}\frac{\partial x}{\partial r} + \frac{\partial w}{\partial y}\frac{\partial y}{\partial r} = \frac{-y}{x^2 + y^2}(\cos \theta) + \frac{x}{x^2 + y^2}(\sin \theta)\\ &= \frac{x \sin \theta - y \cos \theta}{x^2 + y^2} = \frac{r \cos \theta \sin \theta - r \sin \theta \cos \theta}{r^2} = 0.\end{aligned}$$

—CONTINUED—

**19. —CONTINUED—**

Furthermore,

$$\frac{\partial w}{\partial \theta} = \frac{\partial w}{\partial x}\frac{\partial x}{\partial \theta} + \frac{\partial w}{\partial y}\frac{\partial y}{\partial \theta} = \frac{-y}{x^2+y^2}(-r\sin\theta) + \frac{x}{x^2+y^2}(r\cos\theta)$$

$$= \frac{-r\sin\theta(-r\sin\theta) + r\cos\theta(r\cos\theta)}{r^2} = \frac{r^2}{r^2} = 1.$$

(b) Since

$$w = \arctan\frac{y}{x} = \arctan\left(\frac{r\sin\theta}{r\cos\theta}\right) = \arctan(\tan\theta) = \theta + n\pi,$$

we have

$$\frac{\partial w}{\partial r} = 0 \quad \text{and} \quad \frac{\partial w}{\partial \theta} = 1.$$

**25.** $F(x, y, z) = x^2 + y^2 + z^2 - 25$

$F_x(x, y, z) = 2x$

$F_y(x, y, z) = 2y$

$F_z(x, y, z) = 2z$

$$\frac{\partial z}{\partial x} = -\frac{F_x(x, y, z)}{F_z(x, y, z)} = -\frac{2x}{2z} = -\frac{x}{z}$$

$$\frac{\partial z}{\partial y} = -\frac{F_y(x, y, z)}{F_z(x, y, z)} = -\frac{2y}{2z} = -\frac{y}{z}$$

**31.** $F(x, y, z) = e^{xz} + xy$

$F_x(x, y, z) = ze^{xz} + y$

$F_y(x, y, z) = x$

$F_z(x, y, z) = xe^{xz}$

$$\frac{\partial z}{\partial x} = -\frac{F_x(x, y, z)}{F_z(x, y, z)} = -\frac{ze^{xz} + y}{xe^{xz}}$$

$$\frac{\partial z}{\partial y} = -\frac{F_y(x, y, z)}{F_z(x, y, z)} = -\frac{x}{xe^{xz}} = -e^{-xz}$$

**37.** $f(x, y) = \dfrac{xy}{\sqrt{x^2 + y^2}}$

$$f_x(x, y) = \frac{\sqrt{x^2+y^2}(y) - xy\left(\frac{2x}{2\sqrt{x^2+y^2}}\right)}{x^2+y^2} = \frac{y^3}{(x^2+y^2)^{3/2}}$$

$$f_y(x, y) = \frac{\sqrt{x^2+y^2}(x) - xy\left(\frac{2y}{2\sqrt{x^2+y^2}}\right)}{x^2+y^2} = \frac{x^3}{(x^2+y^2)^{3/2}}$$

$$f(tx, ty) = \frac{(tx)(ty)}{\sqrt{(tx)^2 + (ty)^2}} = t\left(\frac{xy}{\sqrt{x^2+y^2}}\right) = tf(x, y)$$

Therefore, the function is homogeneous of degree 1.

$$xf_x(x, y) + yf_y(x, y) = x\left[\frac{y^3}{(x^2+y^2)^{3/2}}\right] + y\left[\frac{x^3}{(x^2+y^2)^{3/2}}\right]$$

$$= \frac{xy(y^2+x^2)}{(x^2+y^2)^{3/2}}$$

$$= \frac{xy}{\sqrt{x^2+y^2}} = 1f(x, y)$$

**41.** From the figure we have

$$\sin\frac{\theta}{2} = \frac{b/2}{x} \implies b = 2x\sin\frac{\theta}{2}$$

$$\cos\frac{\theta}{2} = \frac{h}{x} \implies h = x\cos\frac{\theta}{2}.$$

Therefore, the area of the triangle is

$$A = \frac{1}{2}bh = \frac{1}{2}\left(2x\sin\frac{\theta}{2}\right)\left(x\cos\frac{\theta}{2}\right) = \frac{1}{2}x^2\sin\theta.$$

Differentiating the area function with respect to $t$ and substituting the given information about the triangle yields

$$\frac{dA}{dt} = \frac{\partial A}{\partial x}\frac{dx}{dt} + \frac{\partial A}{\partial \theta}\frac{d\theta}{dt}$$

$$= (x\sin\theta)\frac{dx}{dt} + \left(\frac{1}{2}x^2\cos\theta\right)\frac{d\theta}{dt}$$

$$= 6\sin\frac{\pi}{4}\left(\frac{1}{2}\right) + \left(\frac{1}{2}(6^2)\cos\frac{\pi}{4}\right)\left(\frac{\pi}{90}\right)$$

$$= 5\left(\frac{\sqrt{2}}{2}\right)\left(\frac{1}{2}\right) + \frac{1}{2}(36)\left(\frac{\sqrt{2}}{2}\right)\left(\frac{\pi}{90}\right)$$

$$= \frac{\sqrt{2}}{10}(15 + \pi)\text{ m}^2/\text{hr}.$$

**47.** (a) $x = (v_0 \cos \theta)t = (64 \cos 45°)t = 32\sqrt{2}t$

$y = (v_0 \sin \theta)t - 16t^2 = (64 \sin 45°)t - 16t^2 = 32\sqrt{2}t - 16t^2$

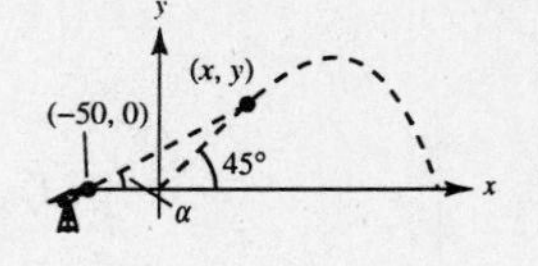


(b) $\tan \alpha = \dfrac{y}{x + 50}$

$$\alpha = \arctan\left(\frac{y}{x + 50}\right) = \arctan\left(\frac{32\sqrt{2}t - 16t^2}{32\sqrt{2}t + 50}\right)$$

(c) $$\frac{d\alpha}{dt} = \frac{1}{1 + \left(\dfrac{32\sqrt{2}t - 16t^2}{32\sqrt{2}t + 50}\right)^2} \cdot \frac{-64\left(8\sqrt{2}t^2 + 25t - 25\sqrt{2}\right)}{\left(32\sqrt{2}t + 50\right)^2}$$

$$= \frac{-16\left(8\sqrt{2}t^2 + 25t - 25\sqrt{2}\right)}{64t^4 - 256\sqrt{2}t^3 + 1024t^2 + 800\sqrt{2}t + 625}$$

(d)

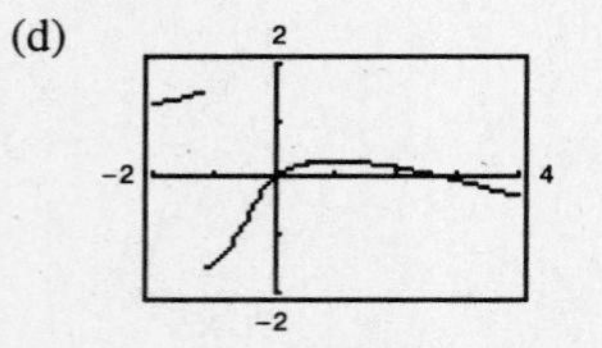


No. The rate of change of $\alpha$ is greatest when the projectile is closest to the camera.

(e) $\dfrac{d\alpha}{dt} = 0$ when

$$8\sqrt{2}t^2 + 25t - 25\sqrt{2} = 0$$

$$t = \frac{-25 + \sqrt{25^2 - 4\left(8\sqrt{2}\right)\left(-25\sqrt{2}\right)}}{2\left(8\sqrt{2}\right)} \approx 0.98 \text{ second.}$$

The projectile is at its maximum height when $dy/dt = 32\sqrt{2} - 32t = 0$ or $t = \sqrt{2} \approx 1.41$ seconds.

## Section 12.6 Directional Derivatives and Gradients

**5.** $g(x, y) = \sqrt{x^2 + y^2}$, $\mathbf{v} = 3\mathbf{i} - 4\mathbf{j}$

The unit vector $\mathbf{u}$ in the direction of $\mathbf{v}$ is

$$\mathbf{u} = \frac{\mathbf{v}}{\|\mathbf{v}\|} = \frac{3}{5}\mathbf{i} - \frac{4}{5}\mathbf{j} = \cos \theta \mathbf{i} + \sin \theta \mathbf{j}.$$

Thus,

$$\cos \theta = \frac{3}{5} \quad \text{and} \quad \sin \theta = -\frac{4}{5}.$$

$$D_{\mathbf{u}}f(x, y) = f_x(x, y) \cos \theta + f_y(x, y) \sin \theta$$

$$= \frac{x}{\sqrt{x^2 + y^2}}\left(\frac{3}{5}\right) + \frac{y}{\sqrt{x^2 + y^2}}\left(-\frac{4}{5}\right)$$

$$= \frac{1}{5\sqrt{x^2 + y^2}}(3x - 4y)$$

Therefore,

$$D_{\mathbf{u}}f(3, 4) = \frac{-7}{25}.$$

Note that $D_{\mathbf{u}}f(x, y) = \nabla f(x, y) \cdot \mathbf{u}$ where

$$\nabla f(x, y) = \frac{x}{\sqrt{x^2 + y^2}}\mathbf{i} + \frac{y}{\sqrt{x^2 + y^2}}\mathbf{j}.$$

**9.** $f(x, y, z) = xy + yz + xz$, $\mathbf{v} = 2\mathbf{i} + \mathbf{j} - \mathbf{k}$

We begin by finding $\nabla f(x, y, z)$ and a unit vector $\mathbf{u}$ in the direction of $\mathbf{v}$.

$$\nabla f(x, y, z) = f_x(x, y, z)\mathbf{i} + f_y(x, y, z)\mathbf{j} + f_z(x, y, z)\mathbf{k}$$
$$= (y + z)\mathbf{i} + (x + z)\mathbf{j} + (x + y)\mathbf{k}$$

and

$$\mathbf{u} = \frac{\mathbf{v}}{\|\mathbf{v}\|} = \frac{\sqrt{6}}{6}(2\mathbf{i} + \mathbf{j} - \mathbf{k}).$$

Therefore,

$$D_{\mathbf{u}}f(x, y, z) = \nabla f(x, y, z) \cdot \mathbf{u}$$

$$= \frac{\sqrt{6}}{6}[2(y + z) + (x + z) - (x + y)$$

$$= \frac{\sqrt{6}}{6}(y + 3z)$$

and

$$D_{\mathbf{u}}f(1, 1, 1) = \frac{4\sqrt{6}}{6} = \frac{2\sqrt{6}}{3}.$$

**17.** $f(x, y) = x^2 + 4y^2, P(3, 1), Q(1, -1)$

A vector in the specified direction is

$$\overrightarrow{PQ} = \mathbf{v} = (1 - 3)\mathbf{i} + (-1 - 1)\mathbf{j} = -2\mathbf{i} - 2\mathbf{j}$$

and a unit vector in this direction is

$$\mathbf{u} = \frac{\mathbf{v}}{\|\mathbf{v}\|} = \frac{-2}{\sqrt{8}}\mathbf{i} - \frac{2}{\sqrt{8}}\mathbf{j} = -\frac{1}{\sqrt{2}}\mathbf{i} - \frac{1}{\sqrt{2}}\mathbf{j}.$$

Since $\nabla f(x, y) = f_x(x, y)\mathbf{i} + f_y(x, y)\mathbf{j} = 2x\mathbf{i} + 8y\mathbf{j}$, the gradient at (3, 1) is

$$\nabla f(3, 1) = 6\mathbf{i} + 8\mathbf{j}.$$

Consequently, at (3, 1) the directional derivative is

$$\begin{aligned} D_{\mathbf{u}}f(3, 1) &= \nabla f(3, 1) \cdot \mathbf{u} \\ &= (6\mathbf{i} + 8\mathbf{j}) \cdot \left(-\frac{\sqrt{2}}{2}\mathbf{i} - \frac{\sqrt{2}}{2}\mathbf{j}\right) \\ &= -3\sqrt{2} - 4\sqrt{2} = -7\sqrt{2}. \end{aligned}$$

**25.** $h(x, y) = x \tan y$

The gradient vector is given by

$$\nabla f(x, y) = f_x(x, y)\mathbf{i} + f_y(x, y)\mathbf{j} = \tan y\mathbf{i} + x\sec^2 y\mathbf{j}.$$

At the point $P = (2, \pi/4)$ the gradient is

$$\nabla f\left(2, \frac{\pi}{4}\right) = \tan\frac{\pi}{4}\mathbf{i} + 2\sec^2\frac{\pi}{4}\mathbf{j} = \mathbf{i} + 4\mathbf{j}.$$

Hence, it follows that the maximum value of the directional derivative at the point $P = (2, \pi/4)$ is

$$\left\|\nabla f\left(2, \frac{\pi}{4}\right)\right\| = \sqrt{17}.$$

**29.** $f(x, y, z) = \sqrt{x^2 + y^2 + z^2}$

The gradient vector is given by

$$\begin{aligned} \nabla f(x, y, z) &= f_x(x, y, z)\mathbf{i} + f_y(x, y, z)\mathbf{j} + f_z(x, y, z)\mathbf{k} \\ &= \frac{x}{\sqrt{x^2 + y^2 + z^2}}\mathbf{i} + \frac{y}{\sqrt{x^2 + y^2 + z^2}}\mathbf{j} + \frac{z}{\sqrt{x^2 + y^2 + z^2}}\mathbf{k} \\ &= \frac{x\mathbf{i} + y\mathbf{j} + z\mathbf{k}}{\sqrt{x^2 + y^2 + z^2}}. \end{aligned}$$

At the point (1, 4, 2) the gradient is

$$\nabla f(1, 4, 2) = \frac{1}{\sqrt{21}}(\mathbf{i} + 4\mathbf{j} + 2\mathbf{k}).$$

Hence, the maximum value of the directional derivative at the point (1, 4, 2) is

$$\|\nabla f(1, 4, 2)\| = \frac{1}{\sqrt{21}}\sqrt{1 + 16 + 4} = 1.$$

**35.** $f(x, y) = 3 - \dfrac{x}{3} - \dfrac{y}{2}$

(a) The directional derivative is

$$\begin{aligned} D_{\mathbf{u}}f(x, y) &= f_x(x, y)\cos\theta + f_y(x, y)\sin\theta \\ &= -\frac{1}{3}\cos\theta - \frac{1}{2}\sin\theta. \end{aligned}$$

For $\theta = 4\pi/3$, $x = 3$, and $y = 2$, we have

$$\begin{aligned} D_{\mathbf{u}}f(3, 2) &= -\frac{1}{3}\cos\frac{4\pi}{3} - \frac{1}{2}\sin\frac{4\pi}{3} \\ &= -\frac{1}{3}\left(-\frac{1}{2}\right) - \frac{1}{2}\left(-\frac{\sqrt{3}}{2}\right) = \frac{2 + 3\sqrt{3}}{12}. \end{aligned}$$

(b) For $\theta = -\pi/6$, $x = 3$, and $y = 2$, we have

$$\begin{aligned} D_{\mathbf{u}}f(3, 2) &= -\frac{1}{3}\cos\left(-\frac{\pi}{6}\right) - \frac{1}{2}\sin\left(-\frac{\pi}{6}\right) \\ &= -\frac{1}{3}\left(\frac{\sqrt{3}}{2}\right) - \frac{1}{2}\left(-\frac{1}{2}\right) = \frac{3 - 2\sqrt{3}}{12}. \end{aligned}$$

**37.** $f(x, y) = 3 - \frac{x}{3} - \frac{y}{2}$

(a) Let **u** be a unit vector in the direction of **v**. Then

$$\mathbf{v} = (-2 - 1)\mathbf{i} + (6 - 2)\mathbf{j} = -3\mathbf{i} + 4\mathbf{j}$$

and

$$\mathbf{u} = \frac{\mathbf{v}}{\|\mathbf{v}\|} = \frac{-3\mathbf{i} + 4\mathbf{j}}{\sqrt{25}} = -\frac{3}{5}\mathbf{i} + \frac{4}{5}\mathbf{j}.$$

At (3, 2),

$$\nabla f(3, 2) = f_x(3, 2)\mathbf{i} + f_y(3, 2)\mathbf{j} = -\frac{1}{3}\mathbf{i} - \frac{1}{2}\mathbf{j}.$$

Therefore, the directional derivative in the direction of **v** is

$$\nabla f(3, 2) \cdot \mathbf{u} = \left(-\frac{1}{3}\right)\left(-\frac{3}{5}\right) + \left(-\frac{1}{2}\right)\left(\frac{4}{5}\right)$$

$$= \frac{1}{5} - \frac{2}{5} = -\frac{1}{5}.$$

(b) Let **u** be a unit vector in the direction of **v**. Then

$$\mathbf{v} = (4 - 3)\mathbf{i} + (5 - 2)\mathbf{j} = \mathbf{i} + 3\mathbf{j}$$

and

$$\mathbf{u} = \frac{\mathbf{v}}{\|\mathbf{v}\|} = \frac{\mathbf{i} + 3\mathbf{j}}{\sqrt{10}} = \frac{\sqrt{10}}{10}\mathbf{i} + \frac{3\sqrt{10}}{10}\mathbf{j}.$$

Therefore, the directional derivative in the direction of **v** is

$$\nabla f(3, 2) \cdot \mathbf{u} = \left(-\frac{1}{3}\right)\left(\frac{\sqrt{10}}{10}\right) + \left(-\frac{1}{2}\right)\left(\frac{3\sqrt{10}}{10}\right)$$

$$= -\frac{11\sqrt{10}}{60}.$$

**39.** $f(x, y) = 3 - \frac{x}{3} - \frac{y}{2}$

The maximum value of the directional derivative is $\|\nabla f(3, 2)\|$.

$$f(x, y) = 3 - \frac{x}{3} - \frac{y}{2}$$

$$\nabla f(x, y) = f_x(x, y)\mathbf{i} + f_y(x, y)\mathbf{j} = -\frac{1}{3}\mathbf{i} - \frac{1}{2}\mathbf{j}$$

Therefore, the maximum value of the directional derivative at (3, 2) is

$$\|\nabla f(3, 2)\| = \sqrt{\frac{1}{9} + \frac{1}{4}} = \frac{\sqrt{13}}{6}.$$

**47.** $f(x, y) = \frac{x}{x^2 + y^2}$

The level curve for $c = \frac{1}{2}$ is given by

$$\frac{x}{x^2 + y^2} = \frac{1}{2}$$

and is shown in the figure. The normal vector to the level curve at $P(1, 1)$ is $\nabla f(1, 1)$.

$$\nabla f(x, y) = f_x(x, y)\mathbf{i} + f_y(x, y)\mathbf{j}$$

$$= \frac{y^2 - x^2}{(x^2 + y^2)^2}\mathbf{i} - \frac{2xy}{(x^2 + y^2)^2}\mathbf{j}$$

$$\nabla f(1, 1) = -\frac{1}{2}\mathbf{j}$$

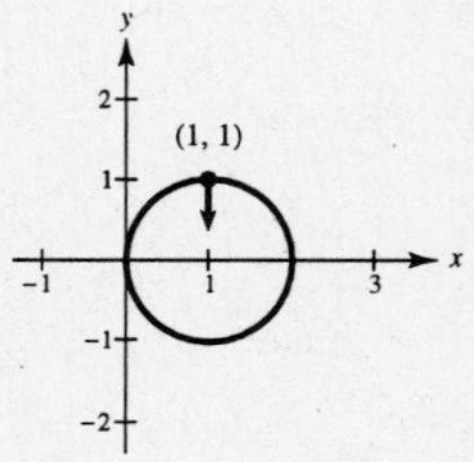

**51.** The ellipse given by the equation $9x^2 + 4y^2 = 40$ corresponds to the level curve with $c = 0$ to the function

$$f(x, y) = 9x^2 + 4y^2 - 40.$$

$\nabla f(x_0, y_0)$ yields a normal vector to the level curve at the point $(x_0, y_0)$. Therefore,

$$\nabla f(x, y) = 18x\mathbf{i} + 8y\mathbf{j}$$

and at $(2, -1)$, a normal vector is

$$\nabla f(2, -1) = 36\mathbf{i} - 8\mathbf{j} = 4(9\mathbf{i} - 2\mathbf{j}).$$

Since, $\|\nabla f(2, -1)\| = 4\sqrt{9^2 + (-2)^2} = 4\sqrt{85}$, a unit normal vector is

$$\frac{\sqrt{85}}{85}(9\mathbf{i} - 2\mathbf{j}).$$

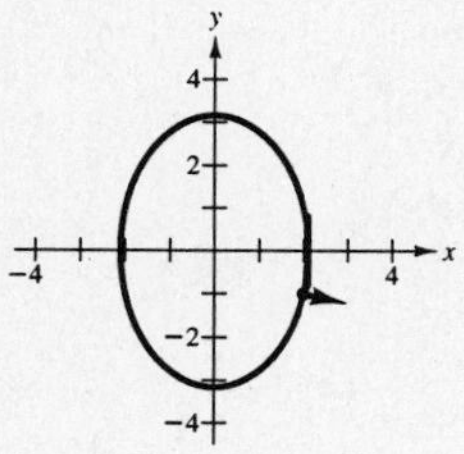

**53.** $T = \dfrac{x}{x^2 + y^2}$

The direction of greatest increase in temperature at $(3, 4)$ will be the direction of the gradient $\nabla T(x, y)$ at that point. Since

$$T_x(x, y) = \frac{(x^2 + y^2)(1) - x(2x)}{(x^2 + y^2)^2} = \frac{y^2 - x^2}{(x^2 + y^2)^2}$$

$$T_y(x, y) = \frac{(x^2 + y^2)(0) - x(2y)}{(x^2 + y^2)^2} = \frac{-2xy}{(x^2 + y^2)^2},$$

the gradient at $(3, 4)$ is

$$\nabla T(3, 4) = T_x(3, 4)\mathbf{i} + T_y(3, 4)\mathbf{j} = \frac{7}{(25)^2}\mathbf{i} - \frac{24}{(25)^2}\mathbf{j} = \frac{1}{625}(7\mathbf{i} - 24\mathbf{j}).$$

**55.** Temperature field: $T(x, y) = 400 - 2x^2 - y^2$

Let the path be represented by the position function

$$\mathbf{r}(t) = x(t)\mathbf{i} + y(t)\mathbf{j}.$$

A tangent vector at each point $(x(t), y(t))$ is given by

$$\mathbf{r}'(t) = \frac{dx}{dt}\mathbf{i} + \frac{dy}{dt}\mathbf{j}.$$

Because the particle seeks maximum temperature increase, the direction of $\mathbf{r}'(t)$ and $\nabla T(x, y) = -4x\mathbf{i} - 2y\mathbf{j}$ are the same at each point of the path. Thus,

$$-4x = k\frac{dx}{dt} \quad \text{and} \quad -2y = k\frac{dy}{dt}$$

where $k$ depends on $t$. By solving each equation for $dt/k$ and equating the results, we have

$$\frac{dx}{-4x} = \frac{dy}{-2y}.$$

The solution of this differential equation is $y^2 = Cx$. Because the particle passes through the point $P(10, 10)$, $C = 10$. Thus, the path of the heat-seeking particle is

$$y^2 = 10x.$$

## Section 12.7 Tangent Planes and Normal Lines

**7.** $z - x \sin y = 4$

Writing the equation for the surface as a function of three variables yields

$$F(x, y, z) = z - x \sin y - 4.$$

A normal vector to the surface, $F(x, y, z) = 0$, at $(x_0, y_0, z_0)$ is given by $\nabla F(x_0, y_0, z_0)$.

$$\begin{aligned}\nabla F(x, y, z) &= F_x(x, y, z)\mathbf{i} + F_y(x, y, z)\mathbf{j} + F_z(x, y, z)\mathbf{k}\\ &= -\sin y\mathbf{i} - x\cos y\mathbf{j} + \mathbf{k}\end{aligned}$$

$$\nabla F\left(6, \frac{\pi}{6}, 7\right) = -\frac{1}{2}\mathbf{i} - 3\sqrt{3}\mathbf{j} + \mathbf{k}$$

Now, the unit normal vector to the surface is

$$\frac{\nabla F(6, \pi/6, 7)}{\|\nabla F(6, \pi/6, 7)\|} = \frac{\sqrt{113}}{113}(-\mathbf{i} - 6\sqrt{3}\mathbf{j} + 2\mathbf{k}).$$

**13.** $f(x, y) = \dfrac{y}{x}$

Begin by writing the equation of the surface as $\dfrac{y}{x} - z = 0$. Then, considering

$$F(x, y, z) = \frac{y}{x} - z,$$

we have

$$F_x(x, y, z) = \frac{-y}{x^2},\ F_y(x, y, z) = \frac{1}{x}, \text{ and } F_z(x, y, z) = -1.$$

At the point (1, 2, 2), the partial derivatives are

$$F_x(1, 2, 2) = -2,\ f_y(1, 2, 2) = 1, \text{ and } F_z(1, 2, 2) = -1.$$

Therefore, the equation of the tangent plane at (1, 2, 2) is

$$\begin{aligned}F_x(1, 2, 2)(x - 1) + F_y(1, 2, 2)(y - 2) + F_z(1, 2, 2)(z - 2) &= 0\\ -2(x - 1) + 1(y - 2) - (z - 2) &= 0\\ -2x + y - z + 2 &= 0.\end{aligned}$$

**23.** $xy^2 + 3x - z^2 = 4$

Let $F(x, y, z) = xy^2 + 3x - z^2 - 4$. Then $\nabla F(2, 1, -2)$ is normal to the tangent plane at $(2, 1, -2)$.

$$\begin{aligned}\nabla F(x, y, z) &= F_x(x, y, z)\mathbf{i} + F_y(x, y, z)\mathbf{j} + F_z(x, y, z)\mathbf{k}\\ &= (y^2 + 3)\mathbf{i} + 2xy\,\mathbf{j} - 2z\mathbf{k}\end{aligned}$$

$$\nabla F(2, 1, -2) = 4\mathbf{i} + 4\mathbf{j} + 4\mathbf{k}$$

Therefore, the equation of the tangent plane is

$$\begin{aligned}4(x - 2) + 4(y - 1) + 4(z + 2) &= 0\\ x + y + z &= 1.\end{aligned}$$

**25.** $x^2 + y^2 + z = 9$

Let $F(x, y, z) = x^2 + y^2 + z - 9$. Then $\nabla F(1, 2, 4)$ is normal to the surface at $(1, 2, 4)$.

$$\begin{aligned}\nabla F(x, y, z) &= F_x(x, y, z)\mathbf{i} + F_y(x, y, z)\mathbf{j} + F_z(x, y, z)\mathbf{k}\\ &= 2x\mathbf{i} + 2y\mathbf{j} + \mathbf{k}\end{aligned}$$

$$\nabla F(1, 2, 4) = 2\mathbf{i} + 4\mathbf{j} + \mathbf{k}$$

Therefore, the equation of the tangent plane is

$$\begin{aligned}2(x - 1) + 4(y - 2) + 1(z - 4) &= 0\\ 2x + 4y + z &= 14.\end{aligned}$$

Since the normal line to the surface at (1, 2, 4) is parallel to $\nabla F(1, 2, 4)$, the direction numbers for the line are 2, 4, and 1. Therefore, symmetric equations for a normal line at (1, 2, 4) are

$$\frac{x - 1}{2} = \frac{y - 2}{4} = \frac{z - 4}{1}.$$

**29.** $z = \arctan \dfrac{y}{x}$

Let $F(x, y, z) = \arctan(y/x) - z$. Then,

$$\nabla F(x, y, z) = F_x(x, y, z)\mathbf{i} + F_y(x, y, z)\mathbf{j} + F_z(x, y, z)\mathbf{k}$$

$$= \frac{-y}{x^2 + y^2}\mathbf{i} + \frac{x}{x^2 + y^2}\mathbf{j} - \mathbf{k}$$

$$\nabla F\left(1, 1, \frac{\pi}{4}\right) = -\frac{1}{2}\mathbf{i} + \frac{1}{2}\mathbf{j} - \mathbf{k} = -\frac{1}{2}(\mathbf{i} - \mathbf{j} + 2\mathbf{k}).$$

Since $\nabla F(1, 1, \pi/4)$ is normal to the surface at the point $(1, 1, \pi/4)$, an equation of the tangent plane is

$$(x - 1) - (y - 1) + 2\left(z - \frac{\pi}{4}\right) = 0$$

$$x - y + 2z = \frac{\pi}{2}$$

and symmetric equations for the normal line are

$$\frac{x - 1}{1} = \frac{y - 1}{-1} = \frac{z - (\pi/4)}{2}.$$

**33.** $x^2 + y^2 = 5,\ z = x$

Let $f(x, y, z) = x^2 + y^2 - 5$ and $g(x, y, z) = x - z$. Then

$$\nabla f(x, y, z) = 2x\mathbf{i} + 2y\mathbf{j} \Rightarrow \nabla f(2, 1, 2) = 4\mathbf{i} + 2\mathbf{j}$$

and

$$\nabla g(x, y, z) = \mathbf{i} - \mathbf{k} \Rightarrow \nabla g(2, 1, 2) = \mathbf{i} - \mathbf{k}.$$

(a) Since $\nabla f$ and $\nabla g$ are each normal to their respective surfaces, the vector $\nabla f \times \nabla g$ will be tangent to both surfaces at the point $(2, 1, 2)$ on the curve of intersection. Therefore, from

$$\nabla f \times \nabla g = \begin{vmatrix} \mathbf{i} & \mathbf{j} & \mathbf{k} \\ 4 & 2 & 0 \\ 1 & 0 & -1 \end{vmatrix} = -2\mathbf{i} + 4\mathbf{j} - 2\mathbf{k}$$

$$= -2(\mathbf{i} - 2\mathbf{j} + \mathbf{k})$$

it follows that direction numbers for the tangent line are 1, $-2$, and 1. Hence, symmetric equations for the tangent line at $(2, 1, 2)$ are

$$\frac{x - 2}{1} = \frac{y - 1}{-2} = \frac{z - 2}{1}.$$

(b) The angle between $\nabla f$ and $\nabla g$ at $(2, 1, 2)$ is such that

$$\cos\theta = \frac{\nabla f \cdot \nabla g}{\|\nabla f\|\|\nabla g\|} = \frac{4 + 0 - 0}{\sqrt{20}\sqrt{2}}$$

$$= \frac{4}{\sqrt{40}} = \frac{4}{2\sqrt{10}} = \frac{\sqrt{10}}{5}.$$

Therefore, the surfaces are **not** orthogonal at the point of intersection.

**41.** If we let

$$F(x, y, z) = 3x^2 + 2y^2 - z - 15$$

then the gradient of $F$ at the point $(2, 2, 5)$ is given by

$$\nabla F(x, y, z) = 6x\mathbf{i} + 4y\mathbf{j} - \mathbf{k}$$

$$\nabla F(2, 2, 5) = 12\mathbf{i} + 8\mathbf{j} - \mathbf{k}$$

Because $\nabla F(2, 2, 5)$ is normal to the tangent plane and $\mathbf{k}$ is normal to the $xy$-plane, it follows that the angle of inclination of the tangent plane is given by

$$\cos\theta = \frac{|\nabla F(2, 2, 5) \cdot \mathbf{k}|}{\|\nabla F(2, 2, 5)\|}$$

$$= \frac{1}{\sqrt{12^2 + 8^2 + (-1)^2}} = \frac{1}{\sqrt{209}}$$

which implies that

$$\theta = \arccos\frac{1}{\sqrt{209}} \approx 86°.$$

**49.** $\dfrac{x^2}{a^2} + \dfrac{y^2}{b^2} + \dfrac{z^2}{c^2} = 1$

We let $F(x, y, z) = \dfrac{x^2}{a^2} + \dfrac{y^2}{b^2} + \dfrac{z^2}{c^2} - 1$. Then

$$\nabla F(x, y, z) = F_x(x, y, z)\mathbf{i} + F_y(x, y, z)\mathbf{j} + F_z(x, y, z)\mathbf{k}$$

$$= \frac{2x}{a^2}\mathbf{i} + \frac{2y}{b^2}\mathbf{j} + \frac{2z}{c^2}\mathbf{k}$$

$$\nabla F(x_0, y_0, z_0) = 2\left[\frac{x_0}{a^2}\mathbf{i} + \frac{y_0}{b^2}\mathbf{j} + \frac{z_0}{c^2}\mathbf{k}\right].$$

Now since $\nabla F(x_0, y_0, z_0)$ is normal to the surface at the point $(x_0, y_0, z_0)$, an equation of the tangent plane is

$$\frac{x_0}{a^2}(x - x_0) + \frac{y_0}{b^2}(y - y_0) + \frac{z_0}{c^2}(z - z_0) = 0$$

$$\left[\frac{x_0x}{a^2} + \frac{y_0y}{b^2} + \frac{z_0z}{c^2}\right] - \left[\frac{x_0^2}{a^2} + \frac{y_0^2}{b^2} + \frac{z_0^2}{c^2}\right] = 0$$

$$\left[\frac{x_0x}{a^2} + \frac{y_0y}{b^2} + \frac{z_0z}{c^2}\right] - 1 = 0$$

$$\frac{x_0x}{a^2} + \frac{y_0y}{b^2} + \frac{z_0z}{c^2} = 1.$$

## Section 12.8 Extrema of Functions of Two Variables

**5.** $f(x, y) = x^2 + y^2 + 2x - 6y + 6$

$$= (x^2 + 2x + \_\_) + (y^2 - 6y + \_\_) + 6 - \_\_ - \_\_$$

$$= (x^2 + 2x + 1) \quad + (y^2 - 6y + 9) \quad + 6 - 1 - 9$$

$$= (x + 1)^2 + (y - 3)^2 - 4 \geq -4.$$

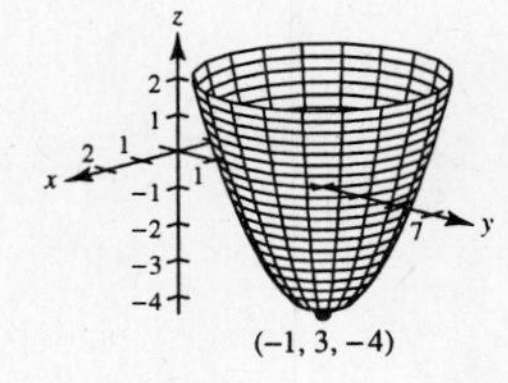


The relative minimum of $f$ is $f(-1, 3) = -4$.

Using partial derivatives to find any critical points and test for relative extrema.

$$f_x(x, y) = 2x + 2 = 0 \text{ when } x = -1$$

$$f_y(x, y) = 2y - 6 = 0 \text{ when } y = 3$$

$$f_{xx}(x, y) = 2$$

$$f_{yy}(x, y) = 2$$

$$f_{xy}(x, y) = 0$$

At the critical point $(-1, 3)$, $f_{xx} > 0$, $f_{yy} > 0$, and $f_{xx}f_{yy} - (f_{xy})^2 > 0$. Therefore, $(-1, 3, -4)$ is a relative minimum.

**13.** $h(x, y) = x^2 - y^2 - 2x - 4y - 4$

Since $h_x(x, y) = 2x - 2 = 2(x - 1) = 0$ when $x = 1$ and $h_y(x, y) = -2y - 4 = -2(y + 2) = 0$ when $y = -2$, there is one critical point, $(1, -2)$. Since

$$h_{xx}(x, y) = 2, \; h_{yy}(x, y) = -2, \text{ and } h_{xy}(x, y) = 0,$$

we have

$$d = h_{xx}(1, -2)h_{yy}(1, -2) - [h_{xy}(1, -2)]^2 = -4 - 0 = -4 < 0.$$

Therefore, by part 3 of Theorem 12.17, we conclude that there is a saddle point at $(1, -2, -1)$.

**17.** $f(x, y) = x^3 - 3xy + y^3$

Since $f_x(x, y) = 3x^2 - 3y$ and $f_y(x, y) = -3x + 3y^2$, any critical points must be the simultaneous solutions of the system of equations

$$3x^2 - 3y = 0 \quad \text{and} \quad 3y^2 - 3x = 0.$$

From the first equation it follows that $y = x^2$. Making this substitution for $y$ in the second equation yields

$$3(x^2)^2 - 3x = 0$$

$$x^4 - x = 0$$

$$x(x^3 - 1) = 0.$$

Therefore, the critical points are $(0, 0)$ and $(1, 1)$. Since

$$f_{xx}(x, y) = 6x, \; f_{yy}(x, y) = 6y, \; f_{xy}(x, y) = -3,$$

we have

$$f_{xx}(0, 0) = 0 \quad \text{and} \quad d = f_{xx}(0, 0)f_{yy}(0, 0) - [f_{xy}(0, 0)]^2 = 0 - 9 < 0.$$

Therefore, by Theorem 12.17, we can conclude that $(0, 0, 0)$ is a saddle point of $f$. At $(1, 1)$ we have

$$f(1, 1) = -1, \; f_{xx}(1, 1) = 6 > 0, \quad \text{and} \quad d = f_{xx}(1, 1)f_{yy}(1, 1) - [f_{xy}(1, 1)]^2 = (6)(6) - 9 > 0.$$

By Theorem 12.17, the point $(1, 1, -1)$ is a relative minimum.

**27.** At the critical point $(x_0, y_0)$,

$$f_{xx}(x_0, y_0)f_{yy}(x_0, y_0) - [f_{xy}(x_0, y_0)]^2 = (-9)(6) - 10^2 < 0.$$

Therefore, $(x_0, y_0, f(x_0, y_0))$ is a saddle point.

**35.** $f(x, y) = 12 - 3x - 2y$ has no critical points. On the line $y = x + 1, 0 \le x \le 1$,

$$f(x, y) = f(x) = 12 - 3x - 2(x + 1) = -5x + 10$$

and the maximum is 10, the minimum is 5. On the line $y = -2x + 4, 1 \le x \le 2$,

$$f(x, y) = f(x) = 12 - 3x - 2(-2x + 4) = x + 4$$

and the maximum is 6, the minimum is 5. On the line $y = -\frac{1}{2}x + 1, 0 \le x \le 2$,

$$f(x, y) = f(x) = 12 - 3x - 2\left(-\tfrac{1}{2}x + 1\right) = -2x + 10$$

and the maximum is 10, the minimum is 6.

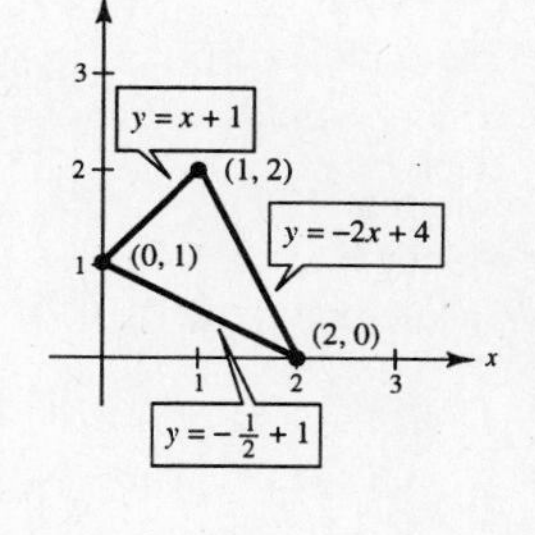


Absolute maximum: 10 at (0, 1)
Absolute minimum: 5 at (1, 2)

**41.** $f(x, y) = x^2 + 2xy + y^2, R = \{(x, y): x^2 + y^2 \le 8\}$

$$\left.\begin{aligned} f_x &= 2x + 2y = 0 \\ f_y &= 2x + 2y = 0 \end{aligned}\right\} y = -x$$

$$f(x, -x) = x^2 - 2x^2 + x^2 = 0$$

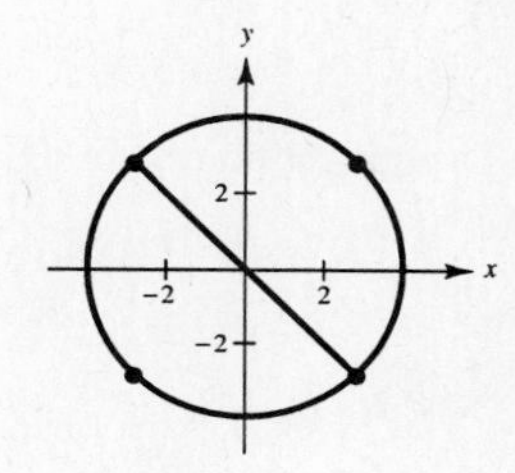


On the boundary $x^2 + y^2 = 8$, we have $y^2 = 8 - x^2$ and $y = \pm\sqrt{8 - x^2}$. Thus

$$f = x^2 \pm 2x\sqrt{8 - x^2} + (8 - x^2) = 8 \pm 2x\sqrt{8 - x^2}$$

$$f' = \pm(x(8 - x^2)^{-1/2}(-2x) + 2(8 - x^2)^{1/2}) = \pm\frac{16 - 4x^2}{\sqrt{8 - x^2}}.$$

Then, $f' = 0$ implies $16 = 4x^2$ or $x = \pm 2$. Thus, the maxima are $f(2, 2) = 16$ and $f(-2, -2) = 16$, and the minima are $f(x, -x) = 0, |x| \le 2$.

**49.** $f(x, y) = x^{2/3} + y^{2/3}$

From the first partial derivatives we have

$$f_x(x, y) = \frac{2}{3}x^{-1/3} = \frac{2}{3\sqrt[3]{x}} \quad \text{and} \quad f_y(x, y) = \frac{2}{3}y^{-1/3} = \frac{2}{3\sqrt[3]{y}}.$$

Since neither first partial derivative exists at (0, 0), it is a critical point. Moreover,

$$f_{xx}(x, y) = -\frac{2}{9x\sqrt[3]{x}} \quad \text{and} \quad f_{yy}(x, y) = -\frac{2}{9x\sqrt[3]{x}}$$

do not exist at (0, 0) and thus the Second-Partials Test fails. Since $x^{2/3} + y^{2/3} \ge 0$ for all $x$ and $y$, it follows that the absolute minimum of $f$ is $f(0, 0) = 0$.

**55.** (a) When $\alpha = 1$ and $\beta = 2$, we have

$$f(x, y) = (x^2 + 2y^2)e^{-(x^2+y^2)}$$

$$f_x(x, y) = -2xe^{-(x^2+y^2)}(x^2 + 2y^2 - 1)$$

$$f_y(x, y) = -2ye^{-(x^2+y^2)}(x^2 + 2y^2 - 2)$$

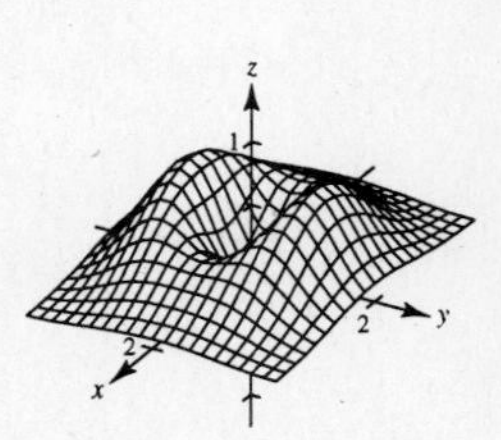


The critical points are (0, 0), $(\pm 1, 0)$, and $(0, \pm 1)$. Using the graph of the function and the critical points, we have the following.

Minimum: (0, 0, 0)
Maxima: $(0, \pm 1, 2e^{-1})$
Saddle points: $(\pm 1, 0, e^{-1})$

**—CONTINUED—**

**55. —CONTINUED—**

(b) When $\alpha = -1$ and $\beta = 2$, we have

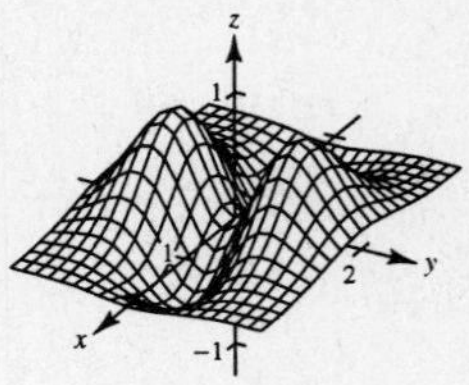

$$f(x, y) = (-x^2 + 2y^2)e^{-(x^2+y^2)}$$

$$f_x(x, y) = 2xe^{-(x^2+y^2)}(x^2 - 2y^2 - 1)$$

$$f_y(x, y) = 2ye^{-(x^2+y^2)}(x^2 - 2y^2 + 2)$$

The critical points are $(0, 0)$, $(\pm 1, 0)$, and $(0, \pm 1)$. Using the graph of the function and the critical points, we have the following.

Minima: $(\pm 1, 0, -e^{-1})$
Maxima: $(0, \pm 1, 2e^{-1})$
Saddle point: $(0, 0, 0)$

(c) $f(x, y) = (\alpha x^2 + \beta y^2)e^{-(x^2+y^2)}$

$$f_x(x, y) = -2xe^{-(x^2+y^2)}(\alpha x^2 + \beta y^2 - \alpha)$$

$$f_y(x, y) = -2ye^{-(x^2+y^2)}(\alpha x^2 + \beta y^2 - \beta)$$

$$f_{xx}(x, y) = 2e^{-(x^2+y^2)}(2\alpha x^4 + 2\beta x^2 y^2 - 5\alpha x^2 - \beta y^2 + \alpha)$$

$$f_{yy}(x, y) = 2e^{-(x^2+y^2)}(2\alpha x^2 y^2 - \alpha x^2 + 2\beta y^4 - 5\beta y^2 + \beta)$$

$$f_{xy}(x, y) = 4xye^{-(x^2+y^2)}(\alpha x^2 + \beta y^2 - \alpha - \beta)$$

The critical points are $(0, 0)$, $(\pm 1, 0)$, and $(0, \pm 1)$. For the case where $\alpha > 0$ and $0 < |\alpha| < \beta$, we have the following.

$$f_{xx}(0, 0) = 2\alpha > 0$$

$$f_{yy}(0, 0) = 2\beta > 0$$

$$f_{xy}(0, 0) = 0$$

$$d = f_{xx}f_{yy} - (f_{xy})^2 > 0$$

Therefore, $(0, 0, 0)$ is a minimum.

$$f_{xx}(0, \pm 1) = 2e^{-1}(-\beta + \alpha) < 0$$

$$f_{yy}(0, \pm 1) = 2e^{-1}(-2\beta) < 0$$

$$f_{xy}(0, 0) = 0$$

$$d = f_{xx}f_{yy} - (f_{xy})^2 > 0$$

Therefore, $(0, \pm 1, \beta e^{-1})$ are maxima.

$$f_{xx}(\pm 1, 0) = (2e^{-1})(-2\alpha) < 0$$

$$f_{yy}(\pm 1, 0) = 2e^{-1}(-\alpha + \beta) > 0$$

$$f_{xy}(0, 0) = 0$$

$$d = f_{xx}f_{yy} - (f_{xy})^2 < 0$$

Therefore, $(\pm 1, 0, \alpha e^{-1})$ are saddle points.

For the case where $\alpha < 0$ and $0 < |\alpha| < \beta$, we have the following.

$$f_{xx}(0, 0) = 2\alpha < 0$$

$$f_{yy}(0, 0) = 2\beta > 0$$

$$f_{xy}(0, 0) = 0$$

$$d = f_{xx}f_{yy} - (f_{xy})^2 < 0$$

Therefore, $(0, 0, 0)$ is a saddle point.

$$f_{xx}(0, \pm 1) = 2e^{-1}(-\beta + \alpha) < 0$$

$$f_{yy}(0, \pm 1) = 2e^{-1}(-2\beta) < 0$$

$$f_{xy}(0, 0) = 0$$

$$d = f_{xx}f_{yy} - (f_{xy})^2 > 0$$

Therefore, $(0, \pm 1, \beta e^{-1})$ are maxima.

$$f_{xx}(\pm 1, 0) = (2e^{-1})(-2\alpha) > 0$$

$$f_{yy}(\pm 1, 0) = 2e^{-1}(-\alpha + \beta) > 0$$

$$f_{xy}(0, 0) = 0$$

$$d = f_{xx}f_{yy} - (f_{xy})^2 > 0$$

Therefore, $(\pm 1, 0, \alpha e^{-1})$ are minima.

## Section 12.9 Applications of Extrema of Functions of Two Variables

**7.** Let $x$, $y$, and $z$ be the numbers and let $s = x^2 + y^2 + z^2$. Since $x + y + z = 30$, it is necessary to minimize

$$s = x^2 + y^2 + (30 - x - y)^2.$$

Setting the first partial derivatives equal to zero yields

$$s_x = 2x - 2(30 - x - y) = 0 \Rightarrow 2x + y = 30$$

$$s_y = 2y - 2(30 - x - y) = 0 \Rightarrow 2x + 4y = 60$$

Subtracting the first equation from the second we obtain the equation

$$3y = 30 \text{ or } y = 10.$$

Substituting this value into the previous equations and solving yields the critical values

$$x = y = z = 10.$$

These values give us the desired minimum, since

$$s_{xx}(10, 10) = 4 > 0 \quad \text{and} \quad s_{xx}(10, 10)s_{yy}(10, 10) - [s_{xy}(10, 10)]^2 = (4)(4) - 2^2 > 0.$$

**15.** $C =$ (cost per mile)(distance from $P$ to $Q$) + (cost per mile)(distance from $Q$ to $R$) + (cost per mile)(distance from $R$ to $S$)

$$= 3k\sqrt{x^2 + 4} + 2k\sqrt{(y - x)^2 + 1} + k(10 - y)$$

Setting the first partials equal to zero yields the system

$$\frac{\partial C}{\partial x} = k\left[\frac{3x}{\sqrt{x^2 + 4}} + \frac{-2(y - x)}{\sqrt{(y - x)^2 + 1}}\right] = 0$$

$$\frac{\partial C}{\partial y} = k\left[\frac{2(y - x)}{\sqrt{(y - x)^2 + 1}} - 1\right] = 0.$$

From the equation $\partial C/\partial y = 0$ we have,

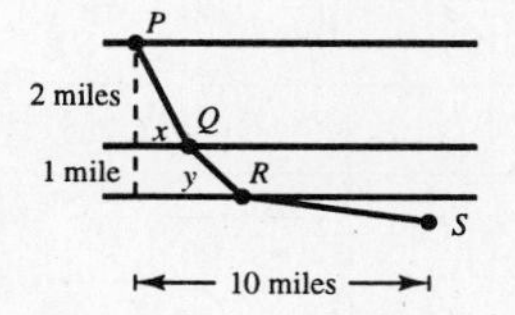


$$\frac{2(y - x)}{\sqrt{(y - x)^2 + 1}} = 1$$

$$2(y - x) = \sqrt{(y - x)^2 + 1}$$

$$4(y - x)^2 = (y - x)^2 + 1$$

$$3(y - x)^2 = 1$$

$$y - x = \pm\frac{1}{\sqrt{3}} \Rightarrow y = x \pm \frac{1}{\sqrt{3}}.$$

Substituting the result with the positive root into the equation $\partial C/\partial x = 0$, we obtain

$$\frac{3x}{\sqrt{x^2 + 4}} = \frac{2(y - x)}{\sqrt{(y - x)^2 + 1}}$$

$$\frac{3x}{\sqrt{x^2 + 4}} = \frac{2/\sqrt{3}}{\sqrt{4/3}}$$

$$3x = \sqrt{x^2 + 4}$$

$$9x^2 = x^2 + 4$$

$$8x^2 = 4.$$

Therefore,

$$x = \frac{1}{\sqrt{2}} \approx 0.707 \text{ mile}$$

$$y = x + \frac{\sqrt{3}}{3} = \frac{3\sqrt{2} + 2\sqrt{3}}{6} \approx 1.284 \text{ miles}.$$

**17.** From the figure observe that the area of a trapezoidal cross section is given by

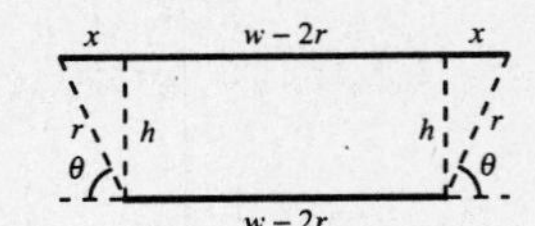


$$A = h\left[\frac{(w - 2r) + [(w - 2r) + 2x]}{2}\right]$$

$$= (w - 2r + x)h$$

where $x = r\cos\theta$ and $h = r\sin\theta$. Substituting these expressions for $x$ and $h$, we have

$$A(r, \theta) = (w - 2r + r\cos\theta)(r\sin\theta)$$

$$= wr\sin\theta - 2r^2\sin\theta + r^2\sin\theta\cos\theta$$

$$A_r(r, \theta) = w\sin\theta - 4r\sin\theta + 2r\sin\theta\cos\theta$$

$$= \sin\theta(w - 4r + 2r\cos\theta) = 0 \implies w = r(4 - 2\cos\theta)$$

and

$$A_\theta(r, \theta) = wr\cos\theta - 2r^2\cos\theta + r^2\cos 2\theta = 0.$$

Substituting the expression for $w$ from $A_r(r, \theta) = 0$ into the equation $A_\theta(r, \theta) = 0$, yields

$$r^2(4 - 2\cos\theta)\cos\theta - 2r^2\cos\theta + r^2(2\cos^2\theta - 1) = 0$$

$$r^2(2\cos\theta - 1) = 0 \text{ or } \cos\theta = \frac{1}{2}.$$

The first partial derivatives are zero when $\theta = \pi/3$ and $r = w/3$. (Ignore the solution $r = \theta = 0$.) Thus, the trapezoid of maximum area occurs when each edge of width $w/3$ is turned up 60° from the horizontal.

**27.** (a)

| $x$ | $y$ | $xy$ | $x^2$ |
|---|---|---|---|
| 0 | 4 | 0 | 0 |
| 1 | 3 | 3 | 1 |
| 1 | 1 | 1 | 1 |
| 2 | 0 | 0 | 4 |
| $\Sigma x_i = 4$ | $\Sigma y_i = 8$ | $\Sigma x_i y_i = 4$ | $\Sigma x_i^2 = 6$ |

By Theorem 12.18, we have

$$a = \frac{n\Sigma x_i y_i - \Sigma x_i \Sigma y_i}{n\Sigma x_i^2 - (\Sigma x_i)^2} = \frac{4(4) - 4(8)}{4(6) - 4^2} = -2$$

$$b = \frac{1}{n}\left(\sum y_i - a\sum x_i\right) = \frac{1}{4}[8 + 2(4)] = 4.$$

Therefore, the least squares regression line is

$$f(x) = -2x + 4.$$

(b) $S = \sum [f(x_i) - y_i]^2$

$= (4 - 4)^2 + (2 - 3)^2 + (2 - 1)^2 + (0 - 0)^2 = 2$

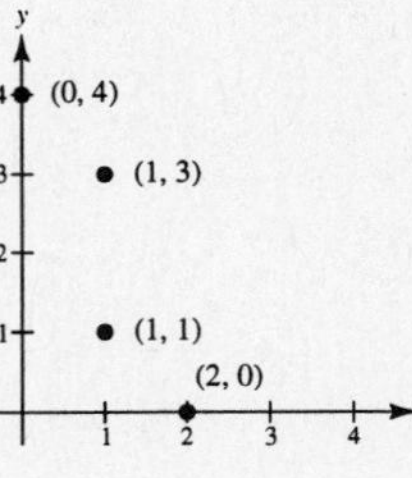


**41.** We are given the points (0, 0), (2, 2), (3, 6), and (4, 12). From Exercise 37, we have that the least squares regression quadratic for the points $(x_1, y_1), (x_2, y_2), \ldots, (x_n, y_n)$ is

$$y = ax^2 + bx + c$$

where $a$, $b$, and $c$ are the solutions to the system

$$a\sum_{i=1}^n x_i^4 + b\sum_{i=1}^n x_i^3 + c\sum_{i=1}^n x_i^2 = \sum_{i=1}^n x_i^2 y_i$$

$$a\sum_{i=1}^n x_i^3 + b\sum_{i=1}^n x_i^2 + c\sum_{i=1}^n x_i = \sum_{i=1}^n x_i y_i$$

$$a\sum_{i=1}^n x_i^2 + b\sum_{i=1}^n x_i + cn = \sum_{i=1}^n y_i.$$

—CONTINUED—

**41. —CONTINUED—**

For the given points, we have

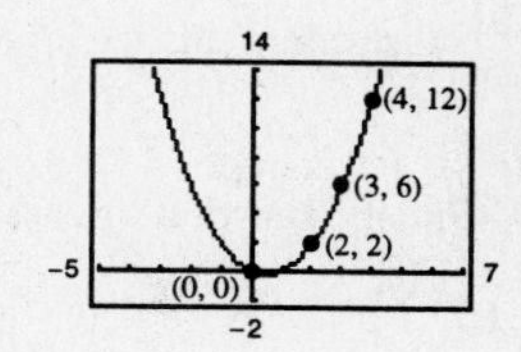


$$\sum x_i = 9, \sum x_i^2 = 29, \sum x_i^3 = 99, \sum x_i^4 = 353, \sum y_i = 20, \sum x_i y_i = 70, \sum x_i^2 y_i = 192.$$

The resulting system of equations is

$$353a + 99b + 29c = 192$$
$$99a + 29b + 9c = 70$$
$$29a + 9b + 4c = 20.$$

Solving this system yields $a = 1$, $b = -1$, and $c = 0$. Therefore, the least squares regression quadratic is

$$y = x^2 - x.$$

# Section 12.10 Lagrange Multipliers

**5.** To minimize $f(x, y) = x^2 - y^2$ subject to the constraint $x - 2y + 6 = 0$, begin by letting

$$g(x, y) = x - 2y + 6.$$

Then, since

$$\nabla f(x, y) = 2x\mathbf{i} - 2y\mathbf{j} \quad \text{and} \quad \lambda \nabla g(x, y) = \lambda(\mathbf{i} - 2\mathbf{j}),$$

we have the following system of equations.

$$2x = \lambda \qquad f_x(x, y) = \lambda g_x(x, y)$$
$$-2y = -2\lambda \qquad f_y(x, y) = \lambda g_y(x, y)$$
$$x - 2y + 6 = 0 \qquad \textit{Constraint}$$

From the first equation we have $\lambda = 2x$. Substituting this result into the second equation, we have $y = 2x$. Substituting this result into the constraint yields

$$x - 2(2x) + 6 = 0$$
$$-3x = -6 \Rightarrow x = 2.$$

Since $y = 2x$, $y = 4$ and the required minimum is

$$f(2, 4) = 2^2 - 4^2 = -12.$$

**13.** To minimize $f(x, y, z) = x^2 + y^2 + z^2$ subject to the constraint $x + y + z - 6 = 0$, begin by letting

$$g(x, y, z) = x + y + z - 6.$$

Since $\nabla f(x, y, z) = 2x\mathbf{i} + 2y\mathbf{j} + 2z\mathbf{k}$ and

$$\lambda \nabla g(x, y, z) = \lambda(\mathbf{i} + \mathbf{j} + \mathbf{k}),$$

we have the following system of equations.

$$2x = \lambda \qquad f_x(x, y, z) = \lambda g_x(x, y, z)$$
$$2y = \lambda \qquad f_y(x, y, z) = \lambda g_y(x, y, z)$$
$$2z = \lambda \qquad f_z(x, y, z) = \lambda g_z(x, y, z)$$
$$x + y + z - 6 = 0 \qquad \textit{Constraint}$$

From the first three equations we have

$$\lambda = 2x = 2y = 2z \quad \text{or} \quad x = y = z.$$

Substituting $x$ for $y$ and $z$ in the constraint produces $3x - 6 = 0$ or $x = 2$. Therefore, $x = 2$, $y = 2$, and $z = 2$ and the required minimum is

$$f(2, 2, 2) = 12.$$

**17.** $f(x, y, z) = xyz$, $x + y + z = 32$, $x - y + z = 0$

In this case there are two constraints which we can denote by $g$ and $h$.

$$g(x, y, z) = x + y + z - 32$$
$$h(x, y, z) = x - y + z$$

Since, $\nabla f(x, y, z) = yz\mathbf{i} + xz\mathbf{j} + xy\mathbf{j}$, $\lambda \nabla g(x, y, z) = \lambda\mathbf{i} + \lambda\mathbf{j} + \lambda\mathbf{k}$, and $\mu \nabla h(x, y, z) = \mu\mathbf{i} - \mu\mathbf{j} + \mu\mathbf{k}$, we have the following system of equations.

$$yz = \lambda + \mu \qquad f_x(x, y, z) = \lambda g_x(x, y, z) + \mu h_x(x, y, z)$$
$$xz = \lambda - \mu \qquad f_y(x, y, z) = \lambda g_y(x, y, z) + \mu h_y(x, y, z)$$
$$xy = \lambda + \mu \qquad f_z(x, y, z) = \lambda g_z(x, y, z) + \mu h_z(x, y, z)$$
$$x + y + z = 32 \qquad \textit{Constraint 1}$$
$$x - y + z = 0 \qquad \textit{Constraint 2}$$

From the first and third equations we have

$$yz = xy \quad \text{or} \quad z = x.$$

**—CONTINUED—**

**17. —CONTINUED—**

Substituting this result into the second constraint, we have

$$y = 2x.$$

Finally, substituting these two results into the first constraint yields

$$4x = 32 \Rightarrow x = 8$$

$$y = 2x = 16$$

$$z = x = 8.$$

Therefore, the maximum value of $f$, subject to the given constraints is

$$f(8, 16, 8) = 8(16)(8) = 1024.$$

**25.** $x + y + z = 1, (2, 1, 1)$

Let $(x, y, z)$ be an arbitrary point in the given plane. Then

$$s = \sqrt{(x-2)^2 + (y-1)^2 + (z-1)^2}$$

represents the distance between $(2, 1, 1)$ and a point in the plane. To simplify our calculations, minimize $s^2$ rather than $s$. With $g(x, y, z) = x + y + z - 1$ as the constraint, we have $\nabla s^2 = 2(x-2)\mathbf{i} + 2(y-1)\mathbf{j} + 2(z-1)\mathbf{k}$ and $\lambda\nabla g(x, y, z) = \lambda\mathbf{i} + \lambda\mathbf{j} + \lambda\mathbf{k}$. Therefore,

$$2(x-2) = \lambda \qquad s_x{}^2(x, y, z) = \lambda g_x(x, y, z)$$

$$2(y-1) = \lambda \qquad s_y{}^2(x, y, z) = \lambda g_y(x, y, z)$$

$$2(z-1) = \lambda \qquad s_z{}^2(x, y, z) - \lambda g_z(x, y, z)$$

$$x + y + z - 1 = 0 \qquad \textit{Constraint.}$$

From the first three equations, we conclude that

$$\lambda = 2(x-2) = 2(y-1) = 2(z-1)$$

or that $x = y + 1$ and $z = y$. Therefore, from the constraint, we have

$$(y+1) + y + y - 1 = 3y = 0 \quad \text{or} \quad y = 0.$$

Thus $x = 1$ and $z = 0$ and the point $(1, 0, 0)$ in the plane $x + y + z = 1$, is closest to the given point $(2, 1, 1)$. The minimum distance is

$$s = \sqrt{(1-2)^2 + (0-1)^2 + (0-1)^2} = \sqrt{3}.$$

**31.** Letting $x$, $y$, and $z$ be the length, width, and height of the solid, respectively, it is necessary to minimize the cost function

$$C(x, y, z) = 5xy + 3(2yz + xy + 2xz)$$

$$= 8xy + 6yz + 6xz$$

subject to the constraint $xyz = 480$. First, write the constraint as

$$g(x, y, z) = xyz - 480.$$

Then, since $\nabla C(x, y, z) = (8y + 6z)\mathbf{i} + (8x + 6z)\mathbf{j} + (6y + 6x)\mathbf{k}$ and $\lambda\nabla g(x, y, z) = \lambda(yz\mathbf{i} + xz\mathbf{j} + xy\mathbf{k})$, we obtain the following system of equations.

$$8y + 6z = \lambda yz \qquad C_x(x, y, z) = \lambda g_x(x, y, z)$$

$$8x + 6z = \lambda xz \qquad C_y(x, y, z) = \lambda g_y(x, y, z)$$

$$6y + 6x = \lambda xy \qquad C_z(x, y, z) = \lambda g_z(x, y, z)$$

$$xyz - 480 = 0 \qquad \textit{Constraint}$$

We now multiply the first equation by $x$, the second by $-y$ and add to obtain

$$6xy - 6yz = 0 \Rightarrow y = x.$$

**—CONTINUED—**

**31. —CONTINUED—**

Next, multiply the first equation by $x$, the third by $-z$ and add to obtain

$$8xy - 6yz = 0 \Rightarrow z = \tfrac{4}{3}x.$$

Finally, substitute these results into the constraint to obtain

$$x(x)\left(\tfrac{4}{3}x\right) = 480$$

$$x^3 = 360$$

$$x = y = \sqrt[3]{360} \text{ and } z = \tfrac{4}{3}\sqrt[3]{360}.$$

**39.** To maximize $P(x, y) = 100x^{0.25}y^{0.75}$ subject to the constraint $48x + 36y = 100{,}000$, begin by letting

$$g(x, y) = 48x + 36y - 100{,}000.$$

Since

$$\nabla P(x, y) = 25x^{-0.75}y^{0.75}\mathbf{i} + 75x^{0.25}y^{-0.25}\mathbf{j} \quad \text{and} \quad \lambda \nabla g(x, y) = \lambda(48\mathbf{i} + 36\mathbf{j}),$$

we have the following system of equations.

$$25x^{-0.75}y^{0.75} = 48\lambda \qquad P_x(x, y) = \lambda g_x(x, y)$$

$$75x^{0.25}y^{-0.25} = 36\lambda \qquad P_y(x, y) = \lambda g_y(x, y)$$

$$48x + 36y - 100{,}000 = 0 \qquad \textit{Constraint}$$

From the first equation we have

$$\left(\frac{y}{x}\right)^{0.75} = \frac{48\lambda}{25},$$

and from the second equation we have

$$\left(\frac{x}{y}\right)^{0.25} = \frac{36\lambda}{75} \text{ or } \left(\frac{y}{x}\right)^{0.25} = \frac{75}{36\lambda}.$$

Therefore,

$$\left(\frac{y}{x}\right)^{0.75}\left(\frac{y}{x}\right)^{0.25} = \left(\frac{48\lambda}{25}\right)\left(\frac{75}{36\lambda}\right)$$

$$\frac{y}{x} = 4$$

$$y = 4x.$$

Substituting this expression for $y$ into the constraint produces

$$48x + 36y = 100{,}000$$

$$192x = 100{,}000.$$

Therefore, $x = \frac{3125}{6}$ and $\frac{6250}{3}$, and the required maximum is

$$P\left(\frac{3125}{6}, \frac{6250}{3}\right) \approx 147{,}314.$$

## Review Exercises for Chapter 12

**11.** $\lim_{(x, y)\to(0, 0)} \frac{-4x^2y}{x^4 + y^2}$

Along the path $y = x^2$ we have

$$\lim_{(x, y)\to(0, 0)} \frac{-4x^2y}{x^4 + y^2} = \lim_{(x, x^2)\to(0, 0)} \frac{-4x^2(x^2)}{x^4 + (x^2)^2}$$
$$= \lim_{(x, x^2)\to(0, 0)} \frac{-4x^4}{2x^4} = -2.$$

Along the path $y = -x^2$ we have

$$\lim_{(x, y)\to(0, 0)} \frac{-4x^2y}{x^4 + y^2} = \lim_{(x, -x^2)\to(0, 0)} \frac{-4x^2(-x^2)}{x^4 + (-x^2)^2}$$
$$= \lim_{(x, -x^2)\to(0, 0)} \frac{4x^4}{2x^4} = 2.$$

Since the limits are not the same along different paths, the limit does not exist. The function is continuous except at $(0, 0)$.

**17.** $g(x, y) = \frac{xy}{x^2 + y^2}$

Using the Quotient Rule we have

$$g_x(x, y) = \frac{(x^2 + y^2)y - xy(2x)}{(x^2 + y^2)^2} = \frac{y(y^2 - x^2)}{(x^2 + y^2)^2}$$

$$g_y(x, y) = \frac{(x^2 + y^2)x - xy(2y)}{(x^2 + y^2)^2} = \frac{x(x^2 - y^2)}{(x^2 + y^2)^2}.$$

**27.** Since $h(x, y) = x \sin y + y \cos x$ we have

$$h_x(x, y) = \sin y - y \sin x \qquad h_{xx}(x, y) = -y \cos x$$
$$h_y(x, y) = x \cos y + \cos x \qquad h(x, y) = -x \sin y.$$

Furthermore,

$$h_{xy}(x, y) = \cos y - \sin x$$

and

$$h_{yx}(x, y) = \cos y - \sin x.$$

**31.** $z = \frac{y}{x^2 + y^2}$

$$\frac{\partial z}{\partial x} = \frac{-2xy}{(x^2 + y^2)^2} = -2y\left[\frac{x}{(x^2 + y^2)^2}\right]$$

$$\frac{\partial^2 z}{\partial x^2} = -2y\left[\frac{(x^2 + y^2)^2 - x(2)(x^2 + y^2)(2x)}{(x^2 + y^2)^4}\right] = \frac{2y(3x^2 - y^2)}{(x^2 + y^2)^3}$$

$$\frac{\partial z}{\partial y} = \frac{(x^2 + y^2) - y(2y)}{(x^2 + y^2)^2} = \frac{x^2 - y^2}{(x^2 + y^2)^2}$$

$$\frac{\partial^2 z}{\partial y^2} = \frac{(x^2 + y^2)^2(-2y) - (x^2 - y^2)(2)(x^2 + y^2)(2y)}{(x^2 + y^2)^4} = \frac{-2y(3x^2 + y^2)}{(x^2 + y^2)^3}$$

Therefore,

$$\frac{\partial^2 z}{\partial x^2} + \frac{\partial^2 z}{\partial y^2} = \frac{2y(3x^2 - y^2)}{(x^2 + y^2)^3} + \frac{(-2y)(3x^2 - y^2)}{(x^2 + y^2)^3} = 0.$$

**33.** $u = x^2 + y^2 + z^2, x = r \cos t, y = r \sin t, z = t$

(a) By the Chain Rule

$$\frac{\partial u}{\partial r} = \frac{\partial u}{\partial x}\frac{\partial x}{\partial r} + \frac{\partial u}{\partial y}\frac{\partial y}{\partial r} + \frac{\partial u}{\partial z}\frac{\partial z}{\partial r}$$
$$= 2x(\cos t) + 2y(\sin t) + 2z(0)$$
$$= 2r \cos t(\cos t) + 2r \sin t(\sin t)$$
$$= 2r(\cos^2 t + \sin^2 t) = 2r$$

and

$$\frac{\partial u}{\partial t} = \frac{\partial u}{\partial x}\frac{\partial x}{\partial t} + \frac{\partial u}{\partial y}\frac{\partial y}{\partial t} + \frac{\partial u}{\partial z}\frac{\partial z}{\partial t}$$
$$= 2x(-r \sin t) + 2y(r \cos t) + 2z(1)$$
$$= 2r \cos t(-r \sin t) + 2r \sin t(r \cos t) + 2t$$
$$= -2r^2 \sin t \cos t + 2r^2 \sin t \cos t + 2t = 2t.$$

—CONTINUED—

**33. —CONTINUED—**

(b) By first substituting the expressions for $x$, $y$, and $z$, we have

$$\begin{aligned} u &= r^2\cos^2 t + r^2\sin^2 t + t^2 \\ &= r^2(\cos^2 t + \sin^2 t) + t^2 = r^2 + t^2. \end{aligned}$$

Therefore,

$$\frac{\partial u}{\partial r} = 2r \quad \text{and} \quad \frac{\partial u}{\partial t} = 2t.$$

**39.** $f(x, y) = \dfrac{y}{x^2 + y^2}$

The gradient is given by

$$\begin{aligned} \nabla f(x, y) &= f_x(x, y)\mathbf{i} + f_y(x, y)\mathbf{j} \\ &= \frac{-2xy}{(x^2 + y^2)^2}\mathbf{i} + \frac{x^2 - y^2}{(x^2 + y^2)^2}\mathbf{j}. \end{aligned}$$

At the point $(1, 1)$ the gradient is

$$\nabla f(1, 1) = \frac{-2(1)(1)}{(1^2 + 1^2)^2}\mathbf{i} + \frac{1^2 - 1^2}{(1^2 + 1^2)^2}\mathbf{j} = -\frac{1}{2}\mathbf{i}.$$

The maximum value of the directional derivative at the point $(1, 1)$ is given by

$$\|\nabla f(1, 1)\| = \frac{1}{2}.$$

**45.** $f(x, y) = -9 + 4x - 6y - x^2 - y^2$

We start by defining a function $F$ and finding its first partial derivatives.

$$\begin{aligned} F(x, y, z) &= z - f(x, y) \\ &= z + 9 - 4x + 6y + x^2 + y^2 \\ F_x(x, y, z) &= -4 + 2x \\ F_y(x, y, z) &= 6 + 2y \\ F_z(x, y, z) &= 1 \end{aligned}$$

Then at $(2, -3, 4)$ we have $F_x(2, -3, 4) = 0$, $F_y(2, -3, 4) = 0$, and $F_z(2, -3, 4) = 1$. Therefore, an equation of the tangent plane at $(2, -3, 4)$ is

$$0(x - 2) + 0(y + 3) + 1(z - 4) = 0 \quad \text{or} \quad z = 4.$$

Furthermore, a normal line at $(2, -3, 4)$ has direction numbers 0, 0, 1 and its parametric equations are

$$x = 2,\ y = -3,\ z = 4 + t.$$

**51.** We begin by setting the first partials of $f$ equal to zero.

$$f(x, y) = xy + \frac{1}{x} + \frac{1}{y}$$

$$f_x(x, y) = y - \frac{1}{x^2} = 0 \implies x^2y = 1$$

$$f_y(x, y) = x - \frac{1}{y^2} = 0 \implies xy^2 = 1$$

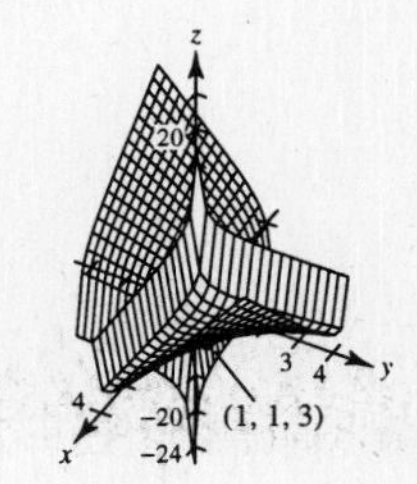


Thus, $x^2y = xy^2$ or $x = y$. Substituting this result into $f_x(x, y) = 0$ yields

$$\begin{aligned} f_x(x, y) &= y - \frac{1}{x^2} \\ &= x - \frac{1}{x^2} \\ &= \frac{x^3 - 1}{x^2} = 0 \implies x = 1. \end{aligned}$$

Therefore, the critical point is $(1, 1)$. We now use the Second Derivative Test and obtain

$$f_{xx}(x, y) = \frac{2}{x^3},\ f_{xy}(x, y) = 1,\ f_{yy}(x, y) = \frac{2}{y^3}.$$

At the critical point $(1, 1)$, we have $f(1, 1) = 3$, $f_{xx}(1, 1) = 2 > 0$, and $f_{xx}(1, 2)f_{yy}(1, 1) - [f_{xy}(1, 1)]^2 = 3 > 0$. Thus, $(1, 1, 3)$ is a relative minimum.

**61.** Using the total differential we have

$$V = \frac{1}{3}\pi r^2 h$$
$$dV = V_r\,dr + V_h\,dh$$
$$= \frac{2}{3}\pi rh\,dr + \frac{1}{3}\pi r^2\,dh.$$

Now, letting $r = 2$, $h = 5$, and $dr = dh = \pm\frac{1}{8}$ we obtain the maximum approximate error.

$$dV = \frac{2}{3}\pi(2)(5)\left(\pm\frac{1}{8}\right) + \frac{1}{3}\pi(2)^2\left(\pm\frac{1}{8}\right)$$
$$= \pm\frac{5}{6}\pi \pm \frac{1}{6}\pi = \pm\pi \text{ in.}^3$$

**69.** To locate the extrema of $w = f(x, y, z) = xy + yz + xz$ subject to the constraint $x + y + z = 1$, begin by letting

$$g(x, y, z) = x + y + z - 1.$$

Since

$$\nabla f(x, y, z) = (y + z)\mathbf{i} + (x + z)\mathbf{j} + (x + y)\mathbf{k} \quad \text{and} \quad \lambda\nabla g(x, y, z) = \lambda(\mathbf{i} + \mathbf{j} + \mathbf{k}),$$

we have the following system of equations.

$$y + z = \lambda \qquad f_x(x, y, z) = \lambda g_x(x, y, z)$$
$$x + z = \lambda \qquad f_y(x, y, z) = \lambda g_y(x, y, z)$$
$$x + y = \lambda \qquad f_z(x, y, z) = \lambda g_z(x, y, z)$$
$$x + y + z - 1 = 0 \qquad \textit{Constraint}$$

From the first three equations we have

$$\lambda = y + z = x + z = x + y \quad \text{or} \quad x = y = z.$$

Substituting $x$ for $y$ and $z$ in the constraint produces $3x - 1 = 0$ or $x = \frac{1}{3}$. Therefore,

$$x = \frac{1}{3},\ y = \frac{1}{3}, \text{ and } z = \frac{1}{3}$$

and the maximum value of the function is

$$f\left(\frac{1}{3}, \frac{1}{3}, \frac{1}{3}\right) = \frac{1}{3}.$$

# CHAPTER 13
# Multiple Integration

# CHAPTER 13
# Multiple Integration

## Section 13.1 Iterated Integrals and Area in the Plane

### Solutions to Selected Odd-Numbered Exercises

**7.** Considering $y$ to be a constant and integrating with respect to $x$ yields

$$\int_{e^y}^{y} y(\ln x)\left(\frac{1}{x}\right) dx = \left[\frac{y(\ln x)^2}{2}\right]_{e^y}^{y}$$

$$= \frac{y}{2}[(\ln y)^2 - (\ln e^y)^2]$$

$$= \frac{y}{2}[(\ln y)^2 - y^2].$$

**13.**
$$\int_1^2 \int_0^4 (x^2 - 2y^2 + 1)\, dx\, dy = \int_1^2 \left[\frac{x^3}{3} - 2xy^2 + x\right]_0^4 dy$$

$$= \int_1^2 \left(\frac{64}{3} - 8y^2 + 4\right) dy$$

$$= \int_1^2 \left(\frac{76}{3} - 8y^2\right) dy$$

$$= \frac{1}{3}\left[76y - 8y^3\right]_1^2$$

$$= \frac{1}{3}[152 - 64 - 76 + 8]$$

$$= \frac{20}{3}$$

**19.**
$$\int_0^{\pi/2} \int_0^{\sin\theta} \theta r\, dr\, d\theta = \int_0^{\pi/2} \left[\frac{\theta r^2}{2}\right]_0^{\sin\theta} d\theta$$

$$= \frac{1}{2}\int_0^{\pi/2} \theta \sin^2\theta\, d\theta$$

$$= \frac{1}{2}\int_0^{\pi/2} \theta\left(\frac{1 - \cos 2\theta}{2}\right) d\theta$$

$$= \frac{1}{4}\int_0^{\pi/2} (\theta - \theta\cos 2\theta)\, d\theta$$

Using integration by parts or a table of integrals yields

$$\int_0^{\pi/2} \int_0^{\sin\theta} \theta r\, dr\, d\theta = \frac{1}{4}\left[\frac{\theta^2}{2} - \left(\frac{1}{4}\cos 2\theta + \frac{\theta}{2}\sin 2\theta\right)\right]_0^{\pi/2}$$

$$= \frac{1}{4}\left(\frac{\pi^2}{8} + \frac{1}{4} - 0 - 0 + \frac{1}{4} + 0\right)$$

$$= \frac{1}{4}\left(\frac{\pi^2}{8} + \frac{1}{2}\right) = \frac{\pi^2}{32} + \frac{1}{8}.$$

**23.**
$$\int_1^{\infty} \int_1^{\infty} \frac{1}{xy}\, dx\, dy = \int_1^{\infty} \lim_{b\to\infty} \int_1^b \frac{1}{xy}\, dx\, dy$$

$$= \int_1^{\infty} \lim_{b\to\infty} \left[\frac{1}{y}\ln x\right]_1^b dy$$

$$= \int_1^{\infty} \lim_{b\to\infty} \frac{1}{y}(\ln b - \ln 1)\, dy$$

$$= \int_1^{\infty} \frac{1}{y}(\infty - 0)\, dy$$

The improper integral diverges.

**35.** $\displaystyle\int_0^1 \int_{y^2}^{\sqrt[3]{y}} dx\, dy$

From the limits of integration, we know that when $0 \le y \le 1$, then $y^2 \le x \le \sqrt[3]{y}$. This implies that the region $R$ is bounded by the curves $x = y^2$ and $x = \sqrt[3]{y}$. Therefore, the region $R$ can be sketched as in the figure. If we interchange the order of integration so that $x$ is the outer variable then $0 \le x \le 1$. Solving for $y$ in the equations $x = y^2$ and $x = \sqrt[3]{y}$, yields $y = \sqrt{x}$ and $y = x^3$. Thus $x^3 \le y \le \sqrt{x}$, and the area of $R$ is given by the iterated integral

$$\int_0^1 \int_{x^2}^{\sqrt{x}} dy\, dx.$$

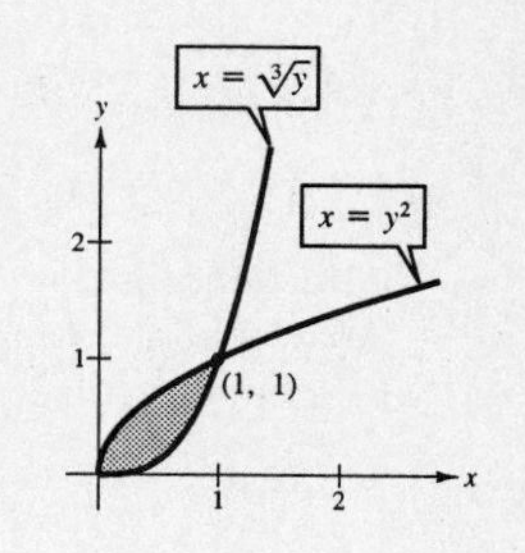


Evaluating each iterated integral, we have

$$\int_0^1 \int_{y^2}^{\sqrt[3]{y}} dx\, dy = \int_0^1 \left(\sqrt[3]{y} - y^2\right) dy = \left[\frac{3}{4}y^{4/3} - \frac{1}{3}y^3\right]_0^1 = \frac{5}{12}$$

$$\int_0^1 \int_{x^3}^{\sqrt{x}} dy\, dx = \int_0^1 \left(\sqrt{x} - x^3\right) dx = \left[\frac{2}{3}x^{3/2} - \frac{1}{4}x^4\right]_0^1 = \frac{5}{12}.$$

**43.** Solving the equation $\sqrt{x} + \sqrt{y} = 2$ for $y$ produces $y = \left(2 - \sqrt{x}\right)^2$. The region bounded by $y = \left(2 - \sqrt{x}\right)^2$, $x = 0$, and $y = 0$ is shown in the figure.

$$A = \int_0^4 \int_0^{(2-\sqrt{x})^2} dy\, dx = \int_0^4 \Big[y\Big]_0^{(2-\sqrt{x})^2} dx$$

$$= \int_0^4 \left(2 - \sqrt{x}\right)^2 dx$$

$$= \int_0^4 \left(4 - 4\sqrt{x} + x\right) dx$$

$$= \left[4x - \frac{8}{3}x\sqrt{x} + \frac{1}{2}x^2\right]_0^4 = \frac{8}{3}$$

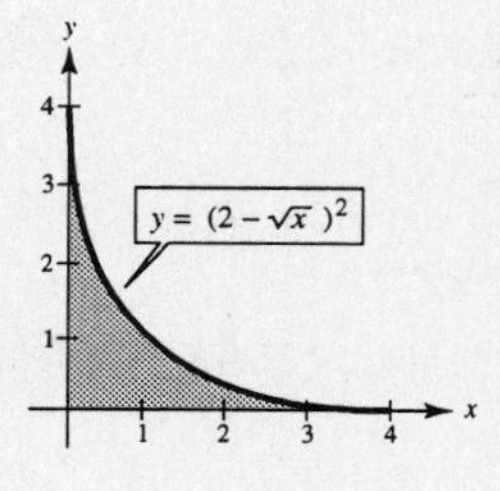


If the order of integration were switched, then

$$A = \int_0^4 \int_0^{(2-\sqrt{y})^2} dx\, dy$$

and the integration steps are similar to those above.

**53.** $\displaystyle\int_0^1 \int_y^1 \sin x^2\, dx\, dy$

Since it is not possible to perform the inner integration, it is necessary to switch the order of integration. From the given limits of integration, it follows that

$y \le x \le 1$ (Inner limits of integration)

which means that the region $R$ is bounded on the left by the line $x = y$ and on the right by $x = 1$. Furthermore, since

$0 \le y \le 1,$ (Outer limits of integration)

it follows that $R$ is bounded below by the $x$-axis as shown in the figure. Now to change the order of integration to $dy\, dx$ observe that the outer limits have the constant bounds $0 \le x \le 1$ and the inner limits have bounds $0 \le y \le x$. Therefore,

$$\int_0^1 \int_y^1 \sin x^2\, dx\, dy = \int_0^1 \int_0^x \sin x^2\, dy\, dx$$

$$= \int_0^1 \Big[y \sin x^2\Big]_0^x dx$$

$$= \int_0^1 x \sin x^2\, dx$$

$$= \frac{1}{2}\Big[-\cos x^2\Big]_0^1 = \frac{1}{2}(1 - \cos 1) \approx 0.230.$$

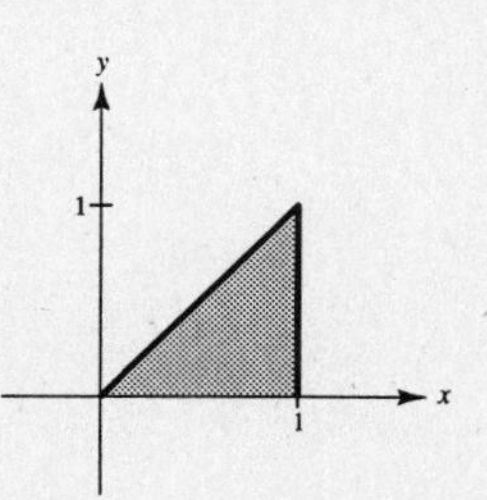

**59.** $\displaystyle\int_0^2 \int_{y^3}^{4\sqrt{2y}} (x^2y - xy^2)\,dx\,dy$

(a) From the given limits of integration, we know that

$y^3 \le x \le 4\sqrt{2y}$ (Inner limits of integration)

which means that the region $R$ is bounded on the left by $x = y^3$ and on the right by $x = 4\sqrt{2y}$. Furthermore, since

$0 \le y \le 2$ (Outer limits of integration)

we have the region $R$ as shown in the figure.

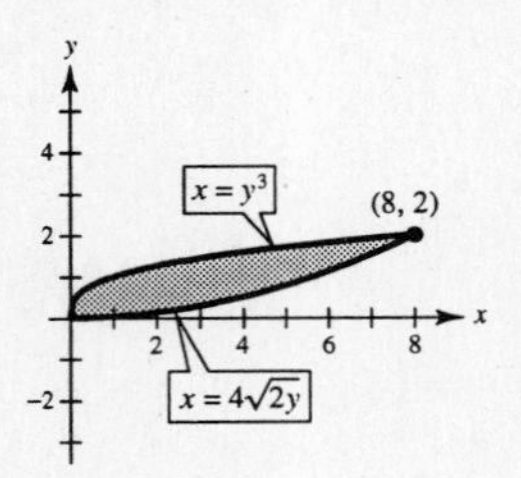


(b) To change the order of integration to $dy\,dx$, place a vertical rectangle in the region. From this you can see that the constant bounds $0 \le x \le 8$ serve as the outer limits of integration. By solving for $y$ in the equations $x = y^3$ and $x = 4\sqrt{2y}$, we have $y = \sqrt[3]{x}$ and $y = x^2/32$, respectively. Therefore, the inner bounds are

$$\frac{x^2}{32} \le y \le \sqrt[3]{x}$$

and the required integral is

$$\int_0^8 \int_{x^2/32}^{\sqrt[3]{x}} (x^2y - xy^2)\,dy\,dx.$$

(c) Using a symbolic integration utility to evaluate both integrals yields

$$\frac{67{,}520}{693}.$$

## Section 13.2 Double Integrals and Volume

**7.** $\displaystyle\int_0^6 \int_{y/2}^{3} (x + y)\,dx\,dy$

From the given limits of integration, it follows that

$\dfrac{y}{2} \le x \le 3$ (Inner limits of integration)

which means that the region $R$ is bounded on the left by $x = y/2$ and on the right by $x = 3$. Furthermore, since

$0 \le y \le 6,$ (Outer limits of integration)

it follows that $R$ is bounded by the $x$-axis as shown in the figure.

$$\int_0^6 \int_{y/2}^{3} (x + y)\,dx\,dy = \int_0^6 \left(\frac{1}{2}x^2 + xy\right)\Big]_{y/2}^{3} dy$$

$$= \int_0^6 \left[\left(\frac{9}{2} + 3y\right) - \left(\frac{y^2}{8} + \frac{y^2}{2}\right)\right] dy$$

$$= \int_0^6 \left(\frac{9}{2} + 3y - \frac{5y^2}{8}\right) dy$$

$$= \left[\frac{9}{2}y + \frac{3}{2}y^2 - \frac{5}{24}y^3\right]_0^6 = 36$$

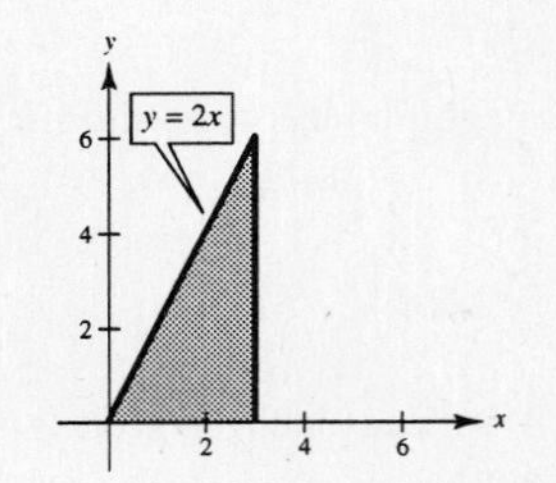

**13.** $R$ is shown in the figure.

$$\int\!\!\int_R \frac{y}{x^2+y^2}\,dx\,dy = \int_0^2\int_{y/2}^{y} \frac{y}{x^2+y^2}\,dx\,dy + \int_2^4\int_{y/2}^{2} \frac{y}{x^2+y^2}\,dx\,dy$$

and

$$\int\!\!\int_R \frac{y}{x^2+y^2}\,dy\,dx = \int_0^2\int_x^{2x} \frac{y}{x^2+y^2}\,dy\,dx$$

$$= \frac{1}{2}\int_0^2 \Big[\ln(x^2+y^2)\Big]_x^{2x}\,dx$$

$$= \frac{1}{2}\int_0^2 [\ln(5x^2) - \ln(2x^2)]\,dx$$

$$= \frac{1}{2}\int_0^2 \ln\frac{5}{2}\,dx$$

$$= \frac{1}{2}\Big[x\ln\frac{5}{2}\Big]_0^2 = \ln\frac{5}{2}$$

**21.** The solid is shown in the figure. By letting $z = 0$, if follows that the base of the solid is the triangle in the $xy$-plane bounded by the graphs of

$2x + 3y = 12$, $x = 0$, and $y = 0$.

If the order of integration is $dy\,dx$, then the bounds of the region are

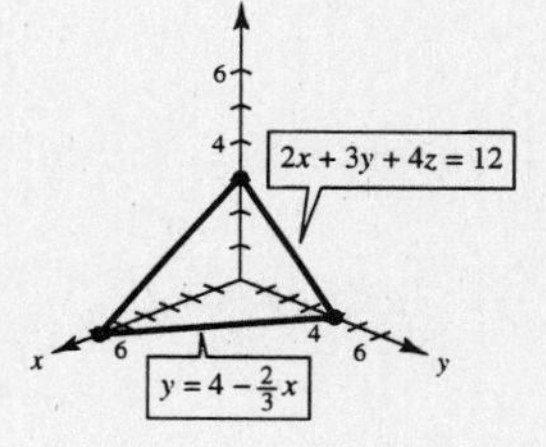


Variable bounds for $y$: $0 \le y \le 4 - \frac{2}{3}x$

Constant bounds for $x$: $0 \le x \le 6$.

Therefore, the volume is

$$V = \int_0^6\int_0^{4-(2/3)x}\left(3 - \frac{1}{2}x - \frac{3}{4}y\right)dy\,dx$$

$$= \int_0^6 \left(3y - \frac{1}{2}xy - \frac{3}{8}y^2\right)\Big]_0^{4-(2/3)x} dx$$

$$= \int_0^6 \left(6 - 2x + \frac{1}{6}x^2\right)dx$$

$$= \left[6x - x^2 + \frac{1}{18}x^3\right]_0^6 = 12.$$

**33.** The figure shows the solid in the first octant. Divide this solid in two equal parts by the plane $y = x$ and find $\frac{1}{2}$ of the total volume. Therefore, integrate the function

$z = \sqrt{1-x^2}$

over the triangle bounded by $y = 0$, $y = x$, and $x = 1$.

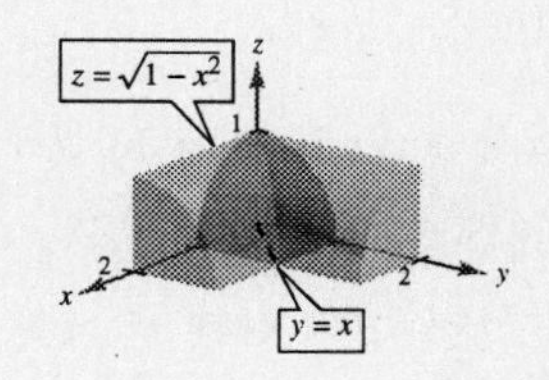


Constant bounds for $x$: $0 \le x \le 1$

Variable bounds for $y$: $0 \le y \le x$

$$V = 2\int_0^1\int_0^x \sqrt{1-x^2}\,dy\,dx$$

$$= 2\int_0^1 x\sqrt{1-x^2}\,dx$$

$$= \left[-\frac{2}{3}(1-x^2)^{3/2}\right]_0^1 = \frac{2}{3}.$$

(Thus, the volume is twice that of Exercise 28.)

**39.** $z = 4 - x^2 - y^2$

Because of the symmetry of the paraboloid, find the volume of the solid in the first octant (one-fourth the total volume). Thus, you integrate over the first-quadrant portion of the circle in the figure.

Constant bounds for $x$: $0 \le x \le 2$

Variable bounds for $y$: $0 \le y \le \sqrt{4 - x^2}$

Using this information and a symbolic integration utility we have

$$V = 4\int_0^2 \int_0^{\sqrt{4-x^2}} (4 - x^2 - y^2)\, dy\, dx = 8\pi.$$

Note that without the aid of a computer or calculator we would have the following

$$\begin{aligned} V &= 4\int_0^2 \int_0^{\sqrt{4-x^2}} (4 - x^2 - y^2)\, dy\, dx \\ &= 4\int_0^2 \left[4y - x^2y - \frac{1}{3}y^3\right]_0^{\sqrt{4-x^2}} dx \\ &= \frac{8}{3}\int_0^2 (4 - x^2)^{3/2}\, dx \\ &= \frac{8}{3}\int_0^{\pi/2} (2\cos\theta)^3(2\cos\theta)\, d\theta \qquad (\text{Let } x = 2\sin\theta.) \\ &= \frac{128}{3}\int_0^{\pi/2} \cos^4\theta\, d\theta = 8\pi. \end{aligned}$$

**47.** $\displaystyle\int_0^{\ln 10} \int_{e^x}^{10} \frac{1}{\ln y}\, dy\, dx$

Since it is not possible to perform the inner integration, it is necessary to switch the order of integration. From the given limits of integration, we know that

$e^x \le y \le 10$ (Inner limits of integration)

and

$0 \le x \le \ln 10.$ (Outer limits of integration)

The region $R$ is shown in the figure. Observe from the figure that if the order of integration is reversed, the outer limits of integration are $1 \le y \le 10$ and the inner limits of integration are $0 \le x \le \ln y$. Therefore,

$$\begin{aligned} \int_0^{\ln 10} \int_{e^x}^{10} \frac{1}{\ln y}\, dy\, dx &= \int_1^{10} \int_0^{\ln y} \frac{1}{\ln y}\, dx\, dy \\ &= \int_1^{10} \left[\frac{x}{\ln y}\right]_0^{\ln y} dy \\ &= \int_1^{10} 1\, dy = \Big[y\Big]_1^{10} = 9. \end{aligned}$$

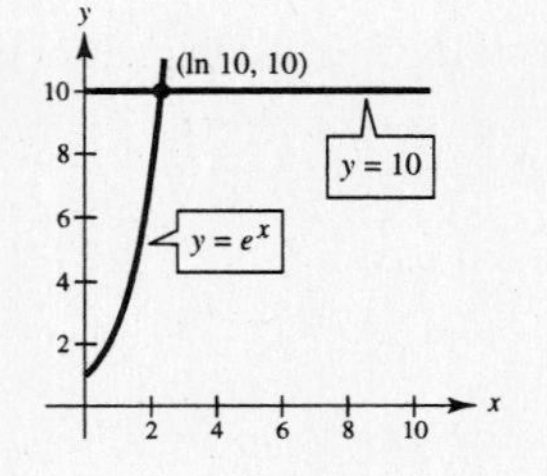


**53.** $f(x, y) = 100x^{0.6}y^{0.4}$, $200 \le x \le 250$, $300 \le y \le 325$

The average value of $f(x, y)$ over the region $R$ is

$$\text{average} = \frac{1}{A}\int_R\int f(x, y)\, dA.$$

The plane region $R$ is a rectangle bounded by $200 \le x \le 250$ and $300 \le y \le 325$. Therefore, its area is

$$A = (250 - 200)(325 - 300) = 1250$$

and the average value of $f$ over the region is

$$\begin{aligned} \text{average} &= \frac{1}{1250}\int_{300}^{325} \int_{200}^{250} 100x^{0.6}y^{0.4}\, dx\, dy \\ &= \frac{1}{1250}\int_{300}^{325} (100y^{0.4})\left[\frac{x^{1.6}}{1.6}\right]_{200}^{250} dy \\ &= \frac{128{,}844.1}{1250}\int_{300}^{325} y^{0.4}\, dy \\ &= 103.075\left[\frac{y^{1.4}}{1.4}\right]_{300}^{325} \approx 25{,}645. \end{aligned}$$

**59.** $f(x, y) = \begin{cases} \frac{1}{27}(9 - x - y), & 0 \le x \le 3, 3 \le y \le 6 \\ 0, & \text{elsewhere} \end{cases}$

First observe that $f(x, y) \ge 0$ for all $(x, y)$.

$$\int_{-\infty}^{\infty}\int_{-\infty}^{\infty} f(x, y)\, dA = \int_0^3\int_3^6 \frac{1}{27}(9 - x - y)\, dy\, dx$$
$$= \int_0^3 \frac{1}{27}\left[9y - xy - \frac{1}{2}y^2\right]_3^6 dx$$
$$= \int_0^3 \left(\frac{1}{2} - \frac{1}{9}x\right) dx$$
$$= \left[\frac{1}{2}x - \frac{1}{18}x^2\right]_0^3 = 1$$

Therefore, $f$ is a joint density function. Finally, we find the specified probability.

$$P(0 \le x \le 1, 4 \le y \le 6) = \int_0^1\int_4^6 \frac{1}{27}(9 - x - y)\, dy\, dx$$
$$= \int_0^1 \frac{1}{27}\left[9y - xy - \frac{1}{2}y^2\right]_4^6 dx$$
$$= \int_0^1 \frac{2}{27}(4 - x)\, dx$$
$$= \frac{2}{27}\left[4x - \frac{1}{2}x^2\right]_0^1 = \frac{7}{27}$$

## Section 13.3 Change of Variables: Polar Coordinates

**5.** $\displaystyle\int_0^{\pi/2}\int_0^{1+\sin\theta} \theta r\, dr\, d\theta$

From the given limits of integration, we know that the inner limits are $0 \le r \le 1 + \sin\theta$ and the outer limits are $0 \le \theta \le \pi/2$. Hence, the region $R$ is the first quadrant portion of the cardioid $r = 1 + \sin\theta$ as shown in the figure.

$$\int_0^{\pi/2}\int_0^{1+\sin\theta} \theta r\, dr\, d\theta = \frac{1}{2}\int_0^{\pi/2}\left[\theta r^2\right]_0^{1+\sin\theta} d\theta$$
$$= \frac{1}{2}\int_0^{\pi/2} (\theta + 2\theta\sin\theta + \theta\sin^2\theta)\, d\theta$$
$$= \frac{1}{2}\int_0^{\pi/2}\left(\frac{3}{2}\theta + 2\theta\sin\theta - \frac{1}{2}\theta\cos 2\theta\right) d\theta = \frac{9}{8} + \frac{3\pi^2}{32}$$

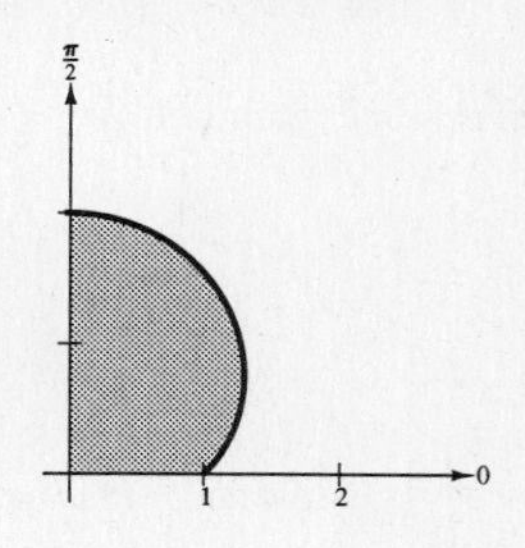


**9.** Using polar axis symmetry, we have

$$A = 2\int_0^{\pi}\int_0^{1+\cos\theta} r\, dr\, d\theta$$
$$= \int_0^{\pi}\left[r^2\right]_0^{1+\cos\theta} d\theta$$
$$= \int_0^{\pi} (1 + 2\cos\theta + \cos^2\theta)\, d\theta$$
$$= \int_0^{\pi}\left(\frac{3}{2} + 2\cos\theta + \frac{1}{2}\cos 2\theta\right) d\theta$$
$$= \left[\frac{3}{2}\theta + 2\sin\theta + \frac{1}{4}\sin 2\theta\right]_0^{\pi} = \frac{3\pi}{2}$$

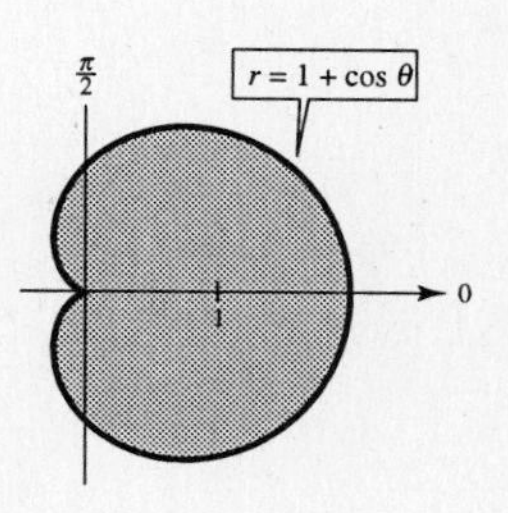

**17.** $\displaystyle\int_0^2\int_0^{\sqrt{2x-x^2}} xy\,dy\,dx$

From the limits of integration it follows that

$$0 \le x \le 2$$
$$0 \le y \le \sqrt{2x-x^2} = \sqrt{1-(x-1)^2}.$$

Therefore, the region of integration is bounded by the semicircle in the first quadrant with radius 1 and center $(1, 0)$ a shown in the figure. In polar coordinates the bounds are

$$0 \le \theta \le \frac{\pi}{2}$$
$$0 \le r \le 2\cos\theta.$$

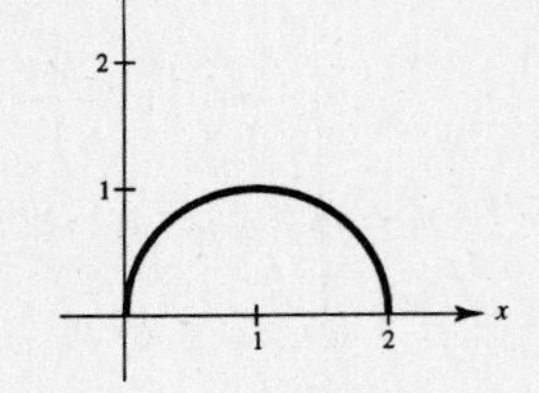


Consequently, the double integral in polar coordinates is

$$\int_0^{\pi/2}\int_0^{2\cos\theta}(r\cos\theta)(r\sin\theta)\,r\,dr\,d\theta = \int_0^{\pi/2}\int_0^{2\cos\theta} r^3\sin\theta\cos\theta\,dr\,d\theta$$
$$= \frac{1}{4}\int_0^{\pi/2}\Big[r^4\sin\theta\cos\theta\Big]_0^{2\cos\theta}d\theta$$
$$= 4\int_0^{\pi/2}\cos^5\theta\sin\theta\,d\theta$$
$$= -\frac{4}{6}\Big[\cos^6\theta\Big]_0^{\pi/2} = \frac{2}{3}.$$

**19.** $\displaystyle\int_0^2\int_0^x \sqrt{x^2+y^2}\,dy\,dx + \int_0^{2\sqrt{2}}\int_0^{\sqrt{8-x^2}}\sqrt{x^2+y^2}\,dy\,dx$

From the figure we can see that $R$ has the bounds

$$0 \le y \le x \qquad 0 \le x \le 2$$
$$0 \le y \le \sqrt{8-x^2} \qquad 2 \le x \le 2\sqrt{2}$$

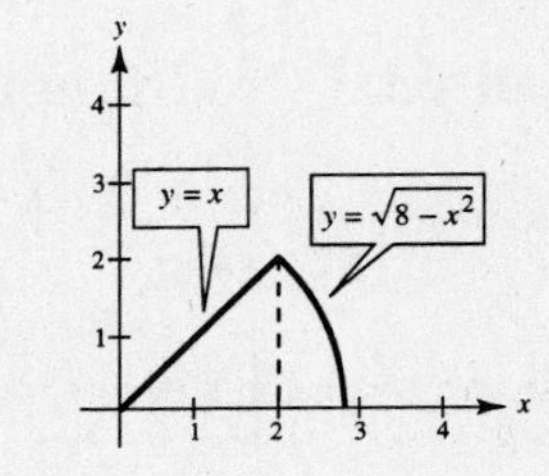


and these bounds form a sector of a circle. In polar coordinates the bounds are

$$0 \le r \le 2\sqrt{2} \quad \text{and} \quad 0 \le \theta \le \frac{\pi}{4}$$

where the integrand is $\sqrt{x^2+y^2} = r$. Consequently, the double integral in polar coordinates is

$$\int_0^{\pi/4}\int_0^{2\sqrt{2}}(r)\overbrace{r\,dr\,d\theta}^{dA} = \int_0^{\pi/4}\Big[\frac{1}{3}r^3\Big]_0^{2\sqrt{2}}d\theta = \frac{16\sqrt{2}}{3}\int_0^{\pi/4}d\theta = \frac{4\sqrt{2}\pi}{3}.$$

**23.** $\displaystyle\int_0^{1/\sqrt{2}}\int_{\sqrt{1-y^2}}^{\sqrt{4-y^2}}\arctan\frac{y}{x}\,dx\,dy + \int_{1/\sqrt{2}}^{\sqrt{2}}\int_y^{\sqrt{4-y^2}}\arctan\frac{y}{x}\,dx\,\cancel{dy}\ \int_0^{\pi/4}\int_1^2 \theta r\,dr\,d\theta$

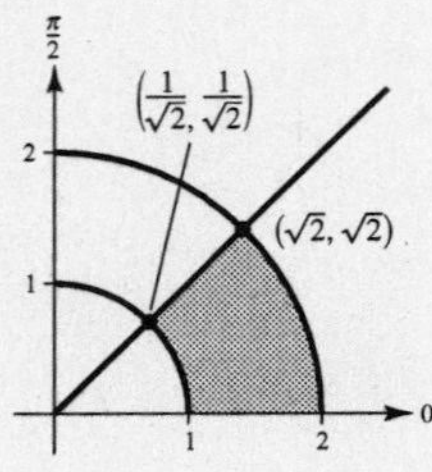

**29.** Writing the equation for the cylinder in polar form we have $r = 4\cos\theta$. We can see from the figure that in polar coordinates $R$ has the bounds

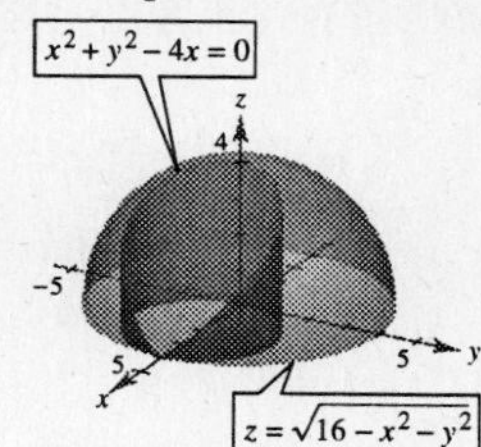


$$0 \le r \le 4\cos\theta \quad \text{and} \quad -\frac{\pi}{2} \le \theta \le \frac{\pi}{2}$$

and $z = \sqrt{16 - x^2 - y^2} = \sqrt{16 - r^2}$. Since the solid is symmetric with respect to the $xz$-plane, the volume $V$ is given by

$$V = 2\int_0^{\pi/2}\int_0^{4\cos\theta} \sqrt{16 - r^2}\, r\, dr\, d\theta = -\frac{2}{3}\int_0^{\pi/2} [(16 - 16\cos^2\theta)^{3/2} - 64]\, d\theta$$

$$= \frac{128}{3}\int_0^{\pi/2} (1 - \sin^3\theta)\, d\theta = \frac{64}{9}(3\pi - 4).$$

**33.** Using symmetry to find the volume of the solid, we have

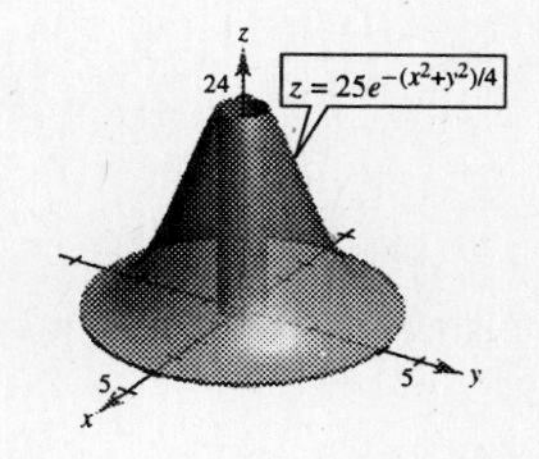


$$V = 4\int_0^{\pi/2}\int_0^4 25e^{-r^2/4} r\, dr\, d\theta$$

$$= -200\int_0^{\pi/2}\int_0^4 e^{-r^2/4}\left(-\frac{r}{2}\right) dr\, d\theta$$

$$= -200\int_0^{\pi/2}\left[e^{r^2/4}\right]_0^4 d\theta$$

$$= 200(1 - e^{-4})\int_0^{\pi/2} d\theta = 100\pi(1 - e^{-4}).$$

Now find the radius $a$ of the hole through the center which is one-tenth the volume of the solid.

$$10\pi(1 - e^{-4}) = 4\int_0^{\pi/2}\int_0^a 25e^{-r^2/4} r\, dr\, d\theta$$

Using the same integration steps as shown above, we have

$$10\pi(1 - e^{-4}) = 100\pi(1 - e^{-a^2/4})$$

$$1 - e^{-4} = 10 - 10e^{-a^2/4}$$

$$e^{-a^2/4} = \frac{9 + e^{-4}}{10}$$

$$-\frac{a^2}{4} = \ln\left(\frac{9 + e^{-4}}{10}\right)$$

$$a^2 = -4\ln\left(\frac{9 + e^{-4}}{10}\right)$$

$$a = \sqrt{-4\ln\left(\frac{9 + e^{-4}}{10}\right)} \approx 0.6429.$$

Therefore, the diameter of the hole is approximately $2(0.6429) = 1.2858$ units.

**45.** Using the symmetry of the density function, we have the following approximation of the population $P$.

$$P = 4\int_0^7\int_0^{\sqrt{49 - x^2}} 4000e^{-0.01(x^2 + y^2)}\, dy\, dx$$

$$= 16{,}000\int_0^{\pi/2}\int_0^7 e^{-0.01r^2} r\, dr\, d\theta$$

$$= 16{,}000\int_0^{\pi/2}\left[-50e^{-0.01r^2}\right]_0^7 d\theta$$

$$= 800{,}000(1 - e^{-0.49})\int_0^{\pi/2} d\theta$$

$$= 800{,}000(1 - e^{-0.49})\frac{\pi}{2} = 400{,}000\pi(1 - e^{-0.49}) \approx 486{,}788$$

# Section 13.4 Center of Mass and Moments of Inertia

**1.** The rectangular lamina is shown in the figure.

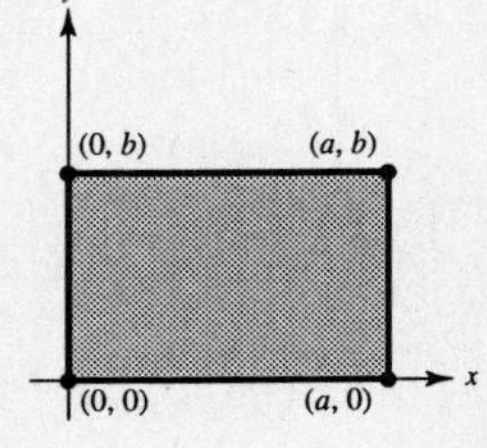


(a) When the lamina has uniform density $\rho = k$, we have the following.

$$m = \int_0^a \int_0^b k\,dy\,dx = kab$$

$$M_x = \int_0^a \int_0^b ky\,dy\,dx = \frac{kab^2}{2}$$

$$M_y = \int_0^a \int_0^b kx\,dy\,dx = \frac{ka^2b}{2}$$

$$\bar{x} = \frac{M_y}{m} = \frac{ka^2b/2}{kab} = \frac{a}{2}$$

$$\bar{y} = \frac{M_x}{m} = \frac{kab^2/2}{kab} = \frac{b}{2}$$

(b) When the lamina has density $\rho = ky$, we have the following.

$$m = \int_0^a \int_0^b ky\,dy\,dx = \frac{kab^2}{2}$$

$$M_x = \int_0^a \int_0^b ky^2\,dy\,dx = \frac{kab^3}{3}$$

$$M_y = \int_0^a \int_0^b kxy\,dy\,dx = \frac{ka^2b^2}{4}$$

$$\bar{x} = \frac{M_y}{m} = \frac{ka^2b^2/4}{kab^2/2} = \frac{a}{2}$$

$$\bar{y} = \frac{M_x}{m} = \frac{kab^3/3}{kab^2/2} = \frac{2}{3}b$$

(c) When the lamina has density $\rho = kx$, we have the following.

$$m = \int_0^a \int_0^b kx\,dy\,dx = \frac{ka^2b}{2}$$

$$M_x = \int_0^a \int_0^b kxy\,dy\,dx = \frac{ka^2b^2}{4}$$

$$M_y = \int_0^a \int_0^b kx^2\,dy\,dx = \frac{ka^3b}{3}$$

$$\bar{x} = \frac{M_y}{m} = \frac{ka^3b/3}{ka^2b/2} = \frac{2a}{3}$$

$$\bar{y} = \frac{M_x}{m} = \frac{ka^2b^2/4}{ka^2b/2} = \frac{b}{2}$$

**7.** The semicircular lamina is shown in the figure.

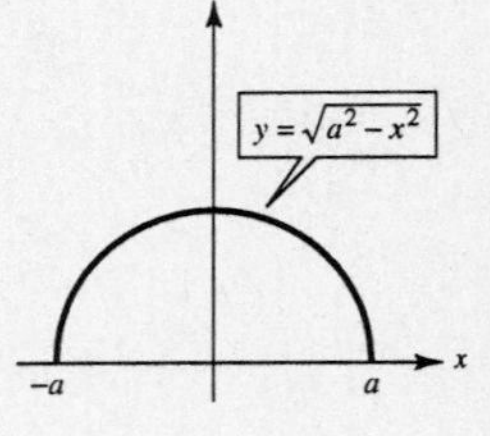


(a) Since the density is uniform ($\rho = k$) and the lamina is symmetric to the $y$-axis, $\bar{x} = 0$.

$$m = \frac{\pi a^2 k}{2}$$

$$M_x = \int_{-a}^a \int_0^{\sqrt{a^2-x^2}} ky\,dy\,dx = \frac{2a^3k}{3}$$

$$\bar{y} = \frac{M_x}{m} = \frac{2a^3k/3}{\pi a^2k/2} = \frac{4a}{3\pi}$$

(b) When the lamina has density $\rho = k(a-y)y$, we have the following.

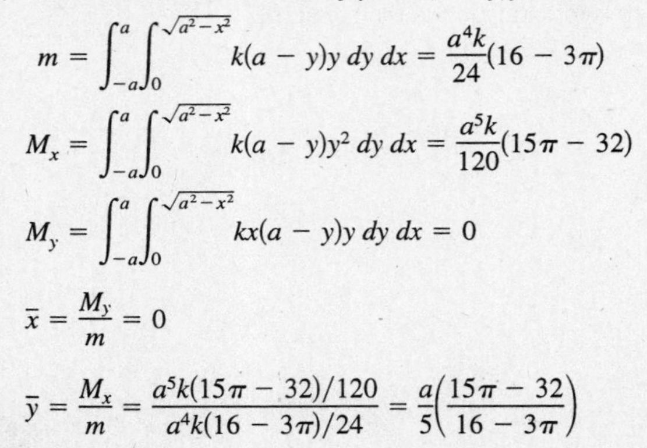

$$m = \int_{-a}^a \int_0^{\sqrt{a^2-x^2}} k(a-y)y\,dy\,dx = \frac{a^4k}{24}(16 - 3\pi)$$

$$M_x = \int_{-a}^a \int_0^{\sqrt{a^2-x^2}} k(a-y)y^2\,dy\,dx = \frac{a^5k}{120}(15\pi - 32)$$

$$M_y = \int_{-a}^a \int_0^{\sqrt{a^2-x^2}} kx(a-y)y\,dy\,dx = 0$$

$$\bar{x} = \frac{M_y}{m} = 0$$

$$\bar{y} = \frac{M_x}{m} = \frac{a^5k(15\pi - 32)/120}{a^4k(16 - 3\pi)/24} = \frac{a}{5}\left(\frac{15\pi - 32}{16 - 3\pi}\right)$$

**15.** By symmetry, we have $\bar{x} = L/2$.

$$m = \int_0^L \int_0^{\sin(\pi x/L)} ky\,dy\,dx$$
$$= \frac{k}{2}\int_0^L \sin^2\left(\frac{\pi x}{L}\right) dx$$
$$= \frac{k}{4}\int_0^L \left[1 - \cos\left(\frac{2\pi x}{L}\right)\right] dx$$
$$= \frac{k}{4}\left[x - \frac{L}{2\pi}\sin\left(\frac{2\pi x}{L}\right)\right]_0^L = \frac{kL}{4}$$
$$M_x = \int_0^L \int_0^{\sin(\pi x/L)} ky^2\,dy\,dx$$
$$= \frac{k}{3}\int_0^L \sin^3\left(\frac{\pi x}{L}\right) dx$$
$$= \frac{k}{3}\int_0^L \left[\sin\left(\frac{\pi x}{L}\right) - \cos^2\left(\frac{\pi x}{L}\right)\sin\left(\frac{\pi x}{L}\right)\right] dx$$
$$= \frac{kL}{3\pi}\left[-\cos\left(\frac{\pi x}{L}\right) + \frac{1}{3}\cos^3\left(\frac{\pi x}{L}\right)\right]_0^L = \frac{4kL}{9\pi}$$
$$\bar{y} = \frac{M_x}{m} = \frac{4kL/9\pi}{kL/4} = \frac{16}{9\pi}$$

**21.** Since the lamina is of uniform density and symmetric to the polar axis (see figure) you have $\bar{y} = 0$.

$$m = \iint_R k\,dA$$
$$= \int_{-\pi/6}^{\pi/6}\int_0^{2\cos 3\theta} kr\,dr\,d\theta \qquad \text{(polar coordinates)}$$
$$= \frac{k\pi}{3}$$
$$M_y = \iint_R kx\,dA$$
$$= k\int_{-\pi/6}^{\pi/6}\int_0^{2\cos 3\theta} (r\cos\theta)r\,dr\,d\theta \qquad \text{(polar coordinates)}$$
$$= \frac{27\sqrt{3}}{40}$$

Therefore,

$$\bar{x} = \frac{M_y}{m} = \frac{27\sqrt{3}k/40}{k\pi/3} = \frac{81\sqrt{3}}{40\pi} \approx 1.12.$$

**29.** Recall that a relationship between polar and rectangular coordinates is $x = r\cos\theta$ and $y = r\sin\theta$.

$$m = \pi a^2$$
$$I_x = \iint_R y^2\rho(x, y)\,dA = \int_0^{2\pi}\int_0^a (r\sin\theta)^2 r\,dr\,d\theta$$
$$= \int_0^{2\pi} \sin^2\theta\left[\frac{r^4}{4}\right]_0^a d\theta = \frac{1}{8}a^4\int_0^{2\pi}(1 - \cos 2\theta)\,d\theta$$
$$= \frac{1}{8}a^4\left[\theta - \frac{1}{2}\sin 2\theta\right]_0^{2\pi} = \frac{1}{4}\pi a^4$$
$$I_y = \iint_R x^2\rho(x, y)\,dA = \int_0^{2\pi}\int_0^a (r\cos\theta)^2 r\,dr\,d\theta$$
$$= \int_0^{2\pi} \cos^2\theta\left[\frac{r^4}{4}\right]_0^a d\theta = \frac{1}{8}a^4\int_0^{2\pi}(1 + \cos 2\theta)\,d\theta$$
$$= \frac{1}{8}a^4\left[\theta + \frac{1}{2}\sin 2\theta\right]_0^{2\pi} = \frac{1}{4}\pi a^4$$
$$I_0 = I_x + I_y = \frac{1}{2}\pi a^4$$
$$\bar{\bar{x}} = \sqrt{\frac{I_x}{m}} = \sqrt{\frac{(\pi a^4)/4}{\pi a^2}} = \frac{a}{2} = \bar{\bar{y}}$$

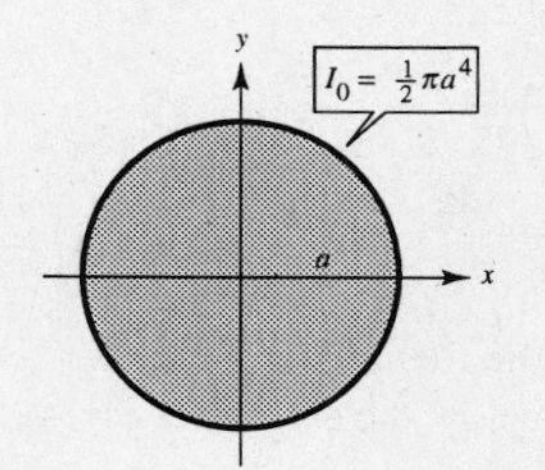

**35.** $y = 4 - x^2, y = 0, x > 0$, and $\rho = kx$

Since $\rho = kx$, we have

$$m = \int_0^2 \int_0^{4-x^2} kx\, dy\, dx$$

$$= k \int_0^2 \left. xy \right]_0^{4-x^2} dx$$

$$= k \int_0^2 x(4 - x^2)\, dx = \left[ -\frac{k}{4}(4 - x^2)^2 \right]_0^2 = 4k.$$

Furthermore,

$$I_x = \int_R\int y^2 \rho\, dA = \int_0^2 \int_0^{4-x^2} kxy^2\, dy\, dx$$

$$= \frac{k}{3}\int_0^2 \left. xy^3 \right]_0^{4-x^2} dx$$

$$= \frac{k}{3}\int_0^2 x(4 - x^2)^3\, dx$$

$$= \left[ -\frac{k}{24}(4 - x^2)^4 \right]_0^2 = \frac{32k}{3}$$

and

$$I_y = \int_R\int x^2 \rho\, dA = \int_0^2 \int_0^{4-x^2} kx^3\, dy\, dx$$

$$= k\int_0^2 \left[ x^3 y \right]_0^{4-x^2} dx$$

$$= k\int_0^2 (4x^3 - x^5)\, dx$$

$$= k\left[ x^4 - \frac{1}{6}x^6 \right]_0^2 = \frac{16k}{3}.$$

Therefore, $I_0 = I_x + I_y = 16k$. Finally,

$$\bar{\bar{x}} = \sqrt{\frac{I_y}{m}} = \sqrt{\frac{16k/3}{4k}} = \frac{2}{\sqrt{3}} = \frac{2\sqrt{3}}{3}$$

$$\bar{\bar{y}} = \sqrt{\frac{I_x}{m}} = \sqrt{\frac{32k/3}{4k}} = \frac{4}{\sqrt{6}} = \frac{2\sqrt{6}}{3}.$$

**41.** $x^2 + y^2 = b^2, x = a\,(a > b)$, and $\rho = k$

$$I = \int_R\int (\text{distance})^2 \text{ mass}$$

$$= \int_{-r}^{r} \int_{-\sqrt{b^2-x^2}}^{\sqrt{b^2-x^2}} (x - a)^2(k)\, dy\, dx$$

$$= 2k\int_0^{\pi}\int_0^b (r\cos\theta - a)^2\, r\, dr\, d\theta \quad \text{(polar coordinates)}$$

$$= \frac{kb^2\pi}{4}(b^2 + 4a^2)$$

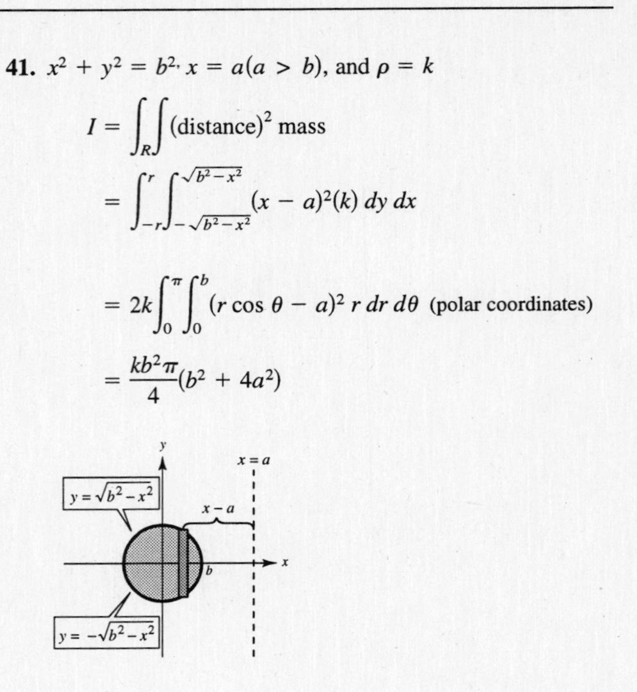


**49.** The gate is shown in the figure. The $y$-coordinate of the centroid of the gate is $\bar{y} = a/2$, the area of the gate is $A = ab$, and the depth of the centroid below the surface of the water is $h = L - a/2$. The moment of inertia of the gate about $\bar{y}$ is

$$I_{\bar{y}} = \int_0^b \int_0^a \left( y - \frac{a}{2} \right)^2 dy\, dx$$

$$= \int_0^b \frac{1}{3}\left[ \left( y - \frac{a}{2} \right)^3 \right]_0^a dx$$

$$= \frac{a^3}{12}\int_0^b dx = \frac{a^3 b}{12}$$

$$y_a = \bar{y} - \frac{I_{\bar{y}}}{hA}$$

$$= \frac{a}{2} - \frac{a^3b/12}{[L - (a/2)](ab)} = \frac{a(3L - 2a)}{3(2L - a)}$$

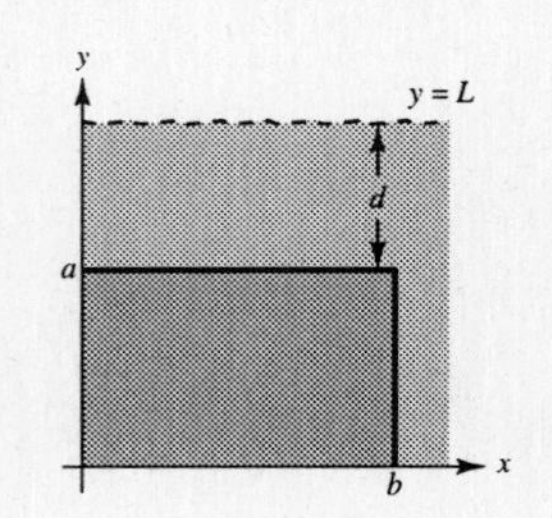

## Section 13.5 Surface Area

**3.** $f(x, y) = 8 + 2x + 2y, R = \{(x, y): x^2 + y^2 \le 4\}$

The first partial derivatives of $f$ are $f_x(x, y) = 2$ and $f_y(x, y) = 2$, and from the formula for surface area it follows that

$$\sqrt{1 + [f_x(x, y)]^2 + [f_y(x, y)]^2} = 3.$$

Therefore, the surface area is given by

$$S = \int_R\int 3\, dA = 3\int_R\int dA = 3(\text{area of } R) = 3(4\pi) = 12\pi.$$

**11.** $f(x, y) = \sqrt{x^2 + y^2}, R = \{(x, y): 0 \le f(x, y) \le 1\}$

Observe that we are to find the surface area of that part of the cone $f(x, y) = \sqrt{x^2 + y^2}$ inside the cylinder $x^2 + y^2 = 1$. Therefore, the region $R$ in the $xy$-plane is a circle of radius 1 centered at the origin. Furthermore, the first partial derivatives of $f$ are

$$f_x(x, y) = \frac{x}{\sqrt{x^2 + y^2}} \quad \text{and} \quad f_y(x, y) = \frac{y}{\sqrt{x^2 + y^2}}$$

and from the formula for surface area, we have

$$\sqrt{1 + [f_x(x, y)]^2 + [f_y(x, y)]^2} = \sqrt{1 + \frac{x^2 + y^2}{x^2 + y^2}} = \sqrt{2}.$$

Therefore, the surface area is given by

$$S = \int_R\int \sqrt{2}\, dA = \sqrt{2}\int_R\int dA = \sqrt{2}(\text{area of } R) = \sqrt{2}\pi.$$

**21.** $f(x, y) = 4 - x^2 - y^2, R = \{(x, y):\ 0 \le f(x, y)\}$

$$0 \le f(x, y) = 4 - x^2 - y^2 \implies x^2 + y^2 \le 4$$

Therefore, we must find the surface area of that part of the paraboloid inside the cylinder $x^2 + y^2 = 4$. The first partial derivatives of $f$ are

$$f_x(x, y) = -2x \quad \text{and} \quad f_y(x, y) = -2y$$

and from the formula for surface area, we have

$$\sqrt{1 + [f_x(x, y)]^2 + [f_y(x, y)]^2} = \sqrt{1 + 4x^2 + 4y^2}.$$

Using the symmetry of the surface, the surface area is

$$\begin{aligned}
S &= 4\int_0^2\int_0^{\sqrt{4-x^2}} \sqrt{1 + 4x^2 + 4y^2}\, dy\, dx \\
&= 4\int_0^{\pi/2}\int_0^2 \sqrt{1 + 4r^2}\, r\, dr\, d\theta \quad \text{(Polar coordinates)} \\
&= 4\left(\frac{1}{8}\right)\left(\frac{2}{3}\right)\int_0^{\pi/2}\left[(1 + 4r^2)^{3/2}\right]_0^2 d\theta \\
&= \frac{1}{3}\int_0^{\pi/2}\left(17\sqrt{17} - 1\right) d\theta \\
&= \frac{17\sqrt{17} - 1}{3}\left[\theta\right]_0^{\pi/2} = \frac{\left(17\sqrt{17} - 1\right)\pi}{6}.
\end{aligned}$$

**31.** $f(x, y) = e^{-x} \sin y$, $R = \{(x, y): x^2 + y^2 \leq 4\}$

The first partial derivatives of $f$ are $f_x(x, y) = -e^{-x} \sin y$ and $f_y(x, y) = e^{-x} \cos y$ and from the formula for surface area, we have

$$\sqrt{1 + [f_x(x, y)]^2 + [f_y(x, y)]^2} = \sqrt{1 + e^{-2x} \sin^2 y + e^{-2x} \cos^2 y} = \sqrt{1 + e^{-2x}}.$$

We integrate over the circle $x^2 + y^2 = 4$.

Constant bounds for $x$: $-2 \leq x \leq 2$

Variable bounds for $y$: $-\sqrt{4 - x^2} \leq y \leq \sqrt{4 - x^2}$

Therefore,

$$S = \int_{-2}^{2} \int_{-\sqrt{4-x^2}}^{\sqrt{4-x^2}} \sqrt{1 + e^{-2x}}\, dy\, dx.$$

**35.** $x^2 + z^2 = 1$, $y^2 + z^2 = 1$

The figure shows the surface in the first octant. We divide this surface into two equal parts by the plane $y = x$ and thus find $\frac{1}{16}$ of the total surface area. Therefore, find the area of the surface $z = \sqrt{1 - x^2}$ over the triangle bounded by $y = 0$, $y = x$, and $x = 1$.

$$\frac{\partial z}{\partial x} = \frac{-x}{\sqrt{1 - x^2}} \quad \text{and} \quad \frac{\partial z}{\partial y} = 0$$

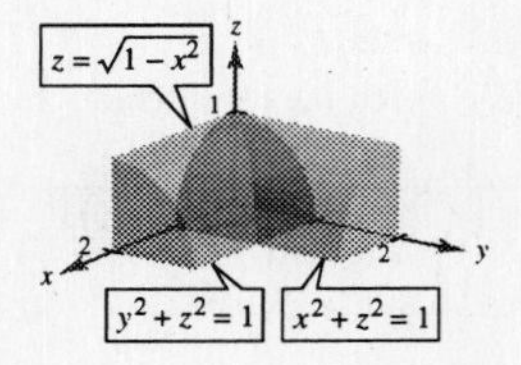


Therefore,

$$S = 16 \int_0^1 \int_0^x \sqrt{1 + \left(\frac{\partial z}{\partial x}\right)^2 + \left(\frac{\partial z}{\partial y}\right)^2}\, dy\, dx$$

$$= 16 \int_0^1 \int_0^x \sqrt{1 + \frac{x^2}{1 - x^2}}\, dy\, dx$$

$$= 16 \int_0^1 \int_0^x \frac{1}{\sqrt{1^2 - x^2}}\, dy\, dx$$

$$= 16 \int_0^1 \frac{x}{\sqrt{1 - x^2}}\, dx = \Big[-16\sqrt{1 - x^2}\Big]_0^1 = 16.$$

**37.** $z = \dfrac{x + y}{5}$ Floor: incline plane

$z = 20 + \dfrac{xy}{100}$, Ceiling

$R = \{(x, y): x^2 + y^2 \leq 50^2, x \geq 0, y \geq 0\}$

(a) $$V = \int_0^{50} \int_0^{\sqrt{50^2 - x^2}} \left[\left(20 + \frac{xy}{100}\right) - \left(\frac{x + y}{5}\right)\right] dy\, dx$$

$$= \int_0^{\pi/2} \int_0^{50} \left[20 + \frac{1}{100} r^2 \sin \theta \cos \theta - \frac{1}{5}(r \sin \theta + r \cos \theta)\right] r\, dr\, d\theta$$

$\approx 30{,}416$ cubic feet

(b) From the equation for the ceiling we have

$$\frac{\partial z}{\partial x} = \frac{y}{100} \quad \text{and} \quad \frac{\partial z}{\partial y} = \frac{x}{100}.$$

$$S = \int_0^{50} \int_0^{\sqrt{50^2 - x^2}} \sqrt{1 + \left(\frac{y}{100}\right)^2 + \left(\frac{x}{100}\right)^2}\, dy\, dx$$

$$= \int_0^{\pi/2} \int_0^{50} \sqrt{1 + \frac{r^2}{100}}\, r\, dr\, d\theta \approx 2082 \text{ square feet.}$$

# Section 13.6 Triple Integrals and Applications

**7.** $$\int_0^9 \int_0^{y/3} \int_0^{\sqrt{y^2-9x^2}} z\,dz\,dx\,dy = \int_0^9 \int_0^{y/3} \left[\frac{1}{2}z^2\right]_0^{\sqrt{y^2-9x^2}} dx\,dy$$
$$= \frac{1}{2}\int_0^9 \int_0^{y/3} (y^2 - 9x^2)\,dx\,dy$$
$$= \frac{1}{2}\int_0^9 \left[xy^2 - 3x^3\right]_0^{y/3} dy$$
$$= \frac{1}{9}\int_0^9 y^3\,dy = \frac{1}{36}\left[y^4\right]_0^9 = \frac{729}{4}$$

**15.** $$\int_0^1 \int_y^1 \int_0^{\sqrt{1-y^2}} dz\,dx\,dy$$

We have

Constant bounds on $y$: $0 \le y \le 1$

Variable bounds on $x$: $y \le x \le 1$

Variable bounds on $z$: $0 \le z \le \sqrt{1-y^2}$.

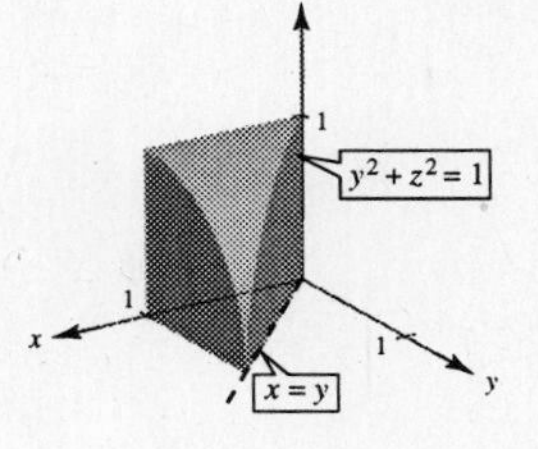


From the upper bound on $z$, we have

$z = \sqrt{1-y^2}$ or $y^2 + z^2 = 1$

a cylinder of radius 1 with the $x$-axis as its axis. Therefore, the triple integral gives the volume of the solid in the first octant bounded by the graphs of $z = \sqrt{1-y^2}$, $z = 0$, $x = y$, and $x = 1$. From the sketch of the solid we observe the following bounds when the order of integration is $dz\,dy\,dx$

Constant bounds on $x$: $0 \le x \le 1$

Variable bounds on $y$: $0 \le y \le x$

Variable bounds on $z$: $0 \le z \le \sqrt{1-y^2}$.

Therefore, the integral is $\displaystyle\int_0^1 \int_0^x \int_0^{\sqrt{1-y^2}} dz\,dy\,dx.$

**23.** In the first octant, we have $0 \le z \le 4 - x^2$, $0 \le y \le 4 - x^2$, and $0 \le x \le 2$. Therefore, the volume is

$$V = \int_0^2 \int_0^{4-x^2} \int_0^{4-x^2} dz\,dy\,dx = \int_0^2 \int_0^{4-x^2} (4 - x^2)\,dy\,dx$$
$$= \int_0^2 (4 - x^2)^2\,dx = \int_0^2 (16 - 8x^2 + x^4)\,dx$$
$$= \left[16x - \frac{8x^3}{3} + \frac{x^5}{5}\right]_0^2 = 32 - \frac{64}{3} + \frac{32}{5} = \frac{256}{15}.$$

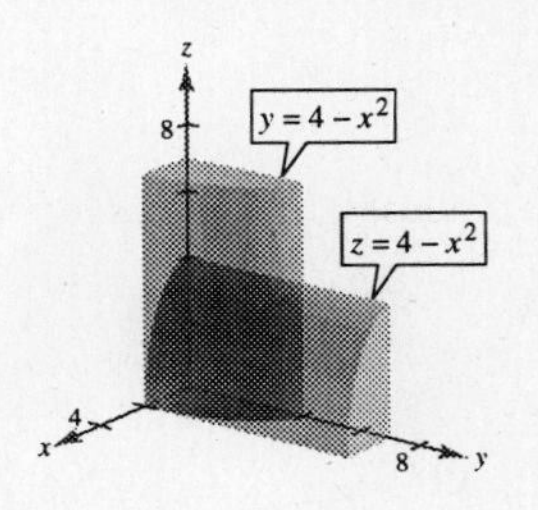


**29.** $x = 0, x = b, y = 0, y = b, z = 0, z = b, \rho(x, y, z) = kxy$

The mass of the cube is

$$m = \int_0^b \int_0^b \int_0^b kxy\,dz\,dy\,dx = k\int_0^b \int_0^b bxy\,dy\,dx$$
$$= kb\int_0^b \left[\frac{xy^2}{2}\right]_0^b dx = \frac{kb^3}{2}\int_0^b x\,dx = \frac{kb^3}{4}\left[x^2\right]_0^b = \frac{kb^5}{4}.$$

—CONTINUED—

**29. —CONTINUED—**

Furthermore,

$$M_{yz} = \int_0^b\int_0^b\int_0^b x(kxy)\,dz\,dy\,dx = k\int_0^b\int_0^b x^2y(b)\,dy\,dx$$

$$= kb\int_0^b\left[\frac{x^2y^2}{2}\right]_0^b dx = \frac{kb^3}{2}\int_0^b x^2\,dx = \frac{kb^3}{6}\Big[x^3\Big]_0^b = \frac{kb^6}{6}.$$

By the symmetry of the cube and of $\rho = kxy$, we have $M_{xz} = M_{yz} = kb^6/6$. Moreover,

$$M_{xy} = \int_0^b\int_0^b\int_0^b z(kxy)\,dz\,dy\,dx = k\int_0^b\int_0^b\left[\frac{xyz^2}{2}\right]_0^b dy\,dx$$

$$= \frac{kb^2}{2}\int_0^b\int_0^b xy\,dy\,dx = \frac{kb^2}{2}\int_0^b\left[\frac{xy^2}{2}\right]_0^b dx$$

$$= \frac{kb^4}{4}\int_0^b x\,dx = \frac{kb^4}{4}\left[\frac{x^2}{2}\right]_0^b = \frac{kb^6}{8}.$$

Finally,

$$\bar{x} = \frac{M_{yz}}{m} = \frac{kb^6/6}{kb^5/4} = \frac{2b}{3}$$

$$\bar{y} = \bar{x} = \frac{2b}{3}$$

$$\bar{z} = \frac{M_{xy}}{m} = \frac{kb^6/8}{kb^5/4} = \frac{b}{2}.$$

**35.** Without loss of generality, position the cone with the vertex at the origin as shown in the figure. Assuming uniform density $\rho(x, y) = k$, the mass of the cone is $m = k(\text{volume}) = \frac{1}{3}k\pi r^2h$. By symmetry we have $\bar{x} = \bar{y} = 0$, and

$$M_{xy} = 4k\int_0^r\int_0^{\sqrt{r^2-x^2}}\int_{h\sqrt{x^2+y^2}/r}^{h} z\,dz\,dy\,dx.$$

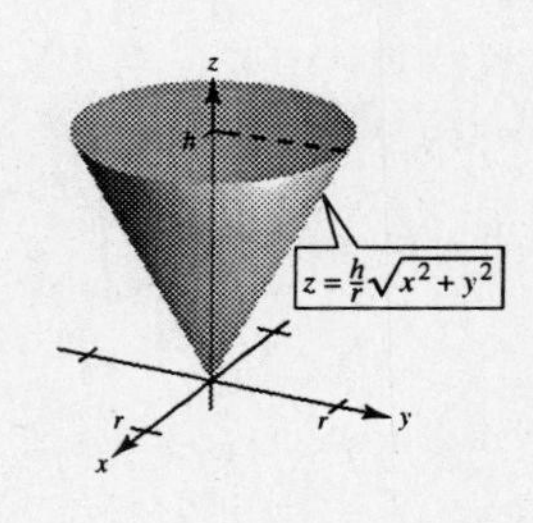


Using a symbolic integration utility to evaluate the triple integral produces

$$M_{xy} = \frac{k\pi r^2h^2}{4}.$$

Therefore, the $z$-coordinate of the centroid of the cone is

$$\bar{z} = \frac{M_{xy}}{m} = \frac{k\pi r^2h/4}{k\pi r^2h/3} = \frac{3h}{4}.$$

**45.** The solid is a right circular of radius $a$, length $L$, and uniform density $\rho(x, y, z) = k$. Thus the mass of the cylinder is

$$m = k(\text{volume}) = k\pi a^2L.$$

Now,

$$I_x = \int_{-a}^{a}\int_{-\sqrt{a^2-x^2}}^{\sqrt{a^2-x^2}}\int_{-L/2}^{L/2}(y^2 + z^2)k\,dy\,dz\,dx$$

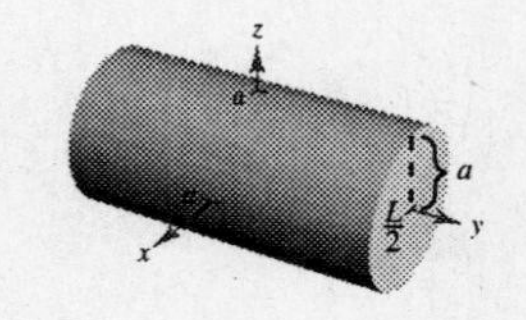


$$= 8k\int_0^a\int_0^{\sqrt{a^2-x^2}}\int_0^{L/2}(y^2 + z^2)\,dy\,dz\,dx \qquad \text{(by symmetry)}$$

$$= \frac{kL}{3}\int_0^a\int_0^{\sqrt{a^2-x^2}}(L^2 + 12z^2)\,dz\,dx$$

$$= \frac{kL}{3}\int_0^a\left[L^2\sqrt{a^2-x^2} + 4(a^2 - x^2)^{3/2}\right]dx$$

—CONTINUED—

**45. —CONTINUED—**

Now let $x = a \sin \theta$. Then $\sqrt{a^2 - x^2} = a \cos \theta$, $dx = a \cos \theta \, d\theta$, and

$$I_x = \frac{ka^2L^3}{3}\int_0^{\pi/2} \cos^2 \theta \, d\theta + \frac{4ka^4L}{3}\int_0^{\pi/2} \cos^4 \theta \, d\theta.$$

By Wallis's Formula we have

$$I_x = \frac{ka^2L^3}{3}\left(\frac{1}{2}\right)\left(\frac{\pi}{2}\right) + \frac{4ka^4L}{3}\left(\frac{1}{2}\right)\left(\frac{3}{4}\right)\left(\frac{\pi}{2}\right)$$

$$= \frac{k\pi a^2L}{12}(L^2 + 3a^2) = \frac{1}{12}m(3a^2 + L^2).$$

By symmetry $I_x = I_z$. To find $I_y$, change only the integrand in the triple integral given above and obtain

$$I_y = 8k\int_0^a\int_0^{\sqrt{a^2-x^2}}\int_0^{L/2} (x^2 + z^2) = dy\,dz\,dx.$$

Proceeding with integrations similar to those given above yields

$$I_y = \frac{k\pi a^4L}{2} = k\pi a^2L\left(\frac{a^2}{2}\right) = \frac{1}{2}ma^2.$$

## Section 13.7 Triple Integrals in Cylindrical and Spherical Coordinates

**3.** $$\int_0^{\pi/2}\int_0^{2\cos^2\theta}\int_0^{4-r^2} r \sin \theta \, dz\,dr\,d\theta = \int_0^{\pi/2}\int_0^{2\cos^2\theta}\left[rz \sin \theta\right]_0^{4-r^2} dr\,d\theta$$

$$= \int_0^{\pi/2}\int_0^{2\cos^2\theta} r(4 - r^2)\sin \theta \, dr\,d\theta$$

$$= \int_0^{\pi/2}\left[-\frac{1}{4}(4 - r^2)^2 \sin \theta\right]_0^{2\cos^2\theta} d\theta$$

$$= \int_0^{\pi/2} (8 \cos^4 \theta - 4 \cos^8 \theta)\sin \theta \, d\theta$$

$$= \left[-\frac{8}{5}\cos^5 \theta + \frac{4}{9}\cos^9 \theta\right]_0^{\pi/2} = \frac{52}{45}$$

**9.** $$\int_0^{\pi/2}\int_0^3\int_0^{e^{-r^2}} r \, dz\,dr\,d\theta$$

We first observe that the triple integral is written in terms of cylindrical coordinates. From the limits of integration we have

Constant bounds on $\theta$: $0 \le \theta \le \pi/2$

Constant bounds on $r$: $0 \le r \le 3$

Variable bounds on $z$: $0 \le z \le e^{-r^2}$.

The limits on $r$ determine a circular cylinder of radius 3 having the $z$-axis as its axis. The limits on $\theta$ and $z$ restrict us to the first-octant portion of the cylinder bounded by the surface

$$z = e^{-r^2} = e^{-(x^2+y^2)}.$$

The solid is shown in the figure. The value of the integral is given by

$$\int_0^{\pi/2}\int_0^3\int_0^{e^{-r^2}} r \, dz\,dr\,d\theta = \int_0^{\pi/2}\int_0^3 re^{-r^2}\,dr\,d\theta$$

$$= -\frac{1}{2}(e^{-9} - 1)\int_0^{\pi/2} d\theta = \frac{\pi}{4}(1 - e^{-9}).$$

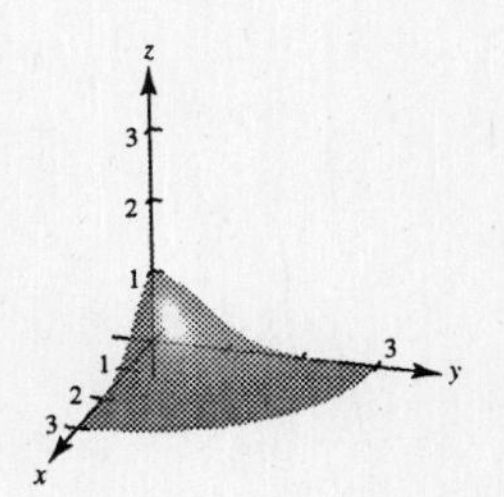

**15.** $\displaystyle\int_{-a}^{a}\int_{-\sqrt{a^2-x^2}}^{\sqrt{a^2-x^2}}\int_{a}^{a+\sqrt{a^2-x^2-y^2}} x\,dz\,dy\,dx$

Observe that the solid $S$ over which we are integrating is the top half of the sphere of radius $a$ and center $(0, 0, a)$. Projecting the solid onto the $xy$-plane forms a circle of radius $a$. Thus, we have

Constant bounds on $\theta$: $0 \le \theta \le 2\pi$

Constant bounds on $r$: $0 \le r \le a$

Variable bounds on $z$: $a \le z \le a + \sqrt{a^2 - x^2 - y^2} = a + \sqrt{a^2 - r^2}$.

Finally, since $x = r\cos\theta$, we obtain the integral in cylindrical coordinates:

$$\int_0^{2\pi}\int_0^{a}\int_0^{a+\sqrt{a^2-r^2}} (r\cos\theta)r\,dz\,dr\,d\theta = \int_0^{2\pi}\int_0^{a}\int_a^{a+\sqrt{a^2-r^2}} r^2\cos\theta\,dz\,dr\,d\theta.$$

To write the integral in spherical coordinates, first write the equation of the sphere in spherical coordinates:

$$z = a + \sqrt{a^2 - x^2 - y^2}$$

$$x^2 + y^2 + (z - a)^2 = a^2$$

$$x^2 + y^2 + z^2 - 2az + a^2 = a^2$$

$$\rho^2 - 2a(\rho\cos\phi) = 0 \qquad (\text{Since } z = \rho\cos\phi)$$

$$\rho = 2a\cos\phi.$$

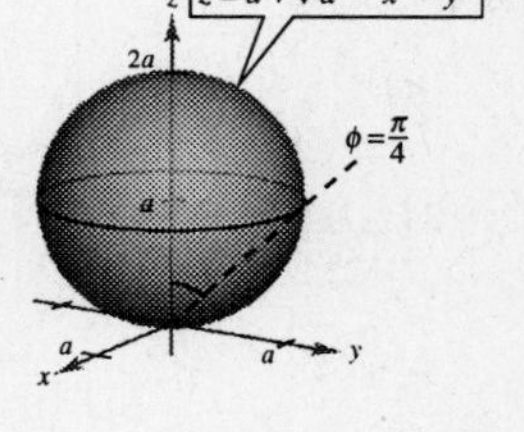


We next write the equation of the plane $z = a$ in spherical coordinates and obtain

$$z = a$$

$$\rho\cos\phi = a$$

$$\rho = a\sec\phi.$$

From the figure it follows that the bounds on $\phi$ are $0 \le \phi \le \pi/4$. Since $x = \rho\sin\phi\cos\theta$, we can write the integral in spherical coordinates.

$$\begin{aligned}\int_0^{\pi/4}\int_0^{2\pi}\int_{a\sec\phi}^{2a\cos\phi} (\rho\sin\phi\cos\theta)(\rho^2\sin\phi)\,d\rho\,d\theta\,d\phi &= \frac{1}{4}\int_0^{\pi/4}\int_0^{2\pi} \sin^2\phi\cos\theta[(2a\cos\phi)^4 - (a\sec\phi)^4]\,d\theta\,d\phi \\ &= \frac{a^4}{4}\int_0^{\pi/4}\Big[(16\sin^2\phi\cos^4\phi - \sin^2\phi\sec^4\phi)\sin\theta\Big]_0^{2\pi}\,d\phi \\ &= \frac{a^4}{4}\int_0^{\pi/4} 0\,d\phi = 0.\end{aligned}$$

**25.** $x^2 + y^2 + z^2 = a^2,\ \left(x - \dfrac{a}{2}\right)^2 + y^2 = \left(\dfrac{a}{2}\right)^2$

The figure shows the sphere and cylinder. Using the cylindrical coordinate system, the equation of the cylinder is given by

$$r = a\cos\theta,\ \left(-\frac{\pi}{2} \le \theta \le \frac{\pi}{2}\right)$$

and the equation of the sphere is given by

$$r^2 + z^2 = a^2 \quad \text{or} \quad z = \pm\sqrt{a^2 - r^2}.$$

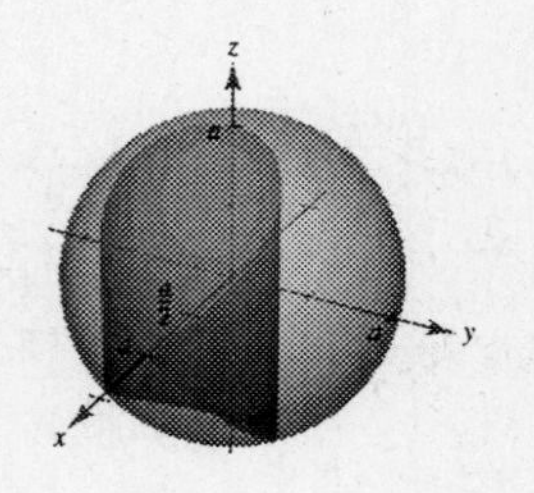


Therefore, the volume is

$$\begin{aligned}V &= \int_{-\pi/2}^{\pi/2}\int_0^{a\cos\theta}\int_{-\sqrt{a^2-r^2}}^{\sqrt{a^2-r^2}} r\,dz\,dr\,d\theta \\ &= 4\int_0^{\pi/2}\int_0^{a\cos\theta} r\sqrt{a^2 - r^2}\,dr\,d\theta \\ &= -\frac{4}{3}\int_0^{\pi/2} [(a^2 - a^2\cos^2\theta)^{3/2} - a^3]\,d\theta \\ &= \frac{4a^3}{3}\int_0^{\pi/2} (1 - \sin^3\theta)\,d\theta = \frac{4a^3}{3}\left(\frac{\pi}{2} - \frac{2}{3}\right).\end{aligned}$$

**33.** Without loss of generality, position the hemisphere with its center at the origin and base on the $xy$-coordinate plane. Then, by symmetry, $\bar{x} = \bar{y} = 0$. Since the solid has uniform density $k$, the mass is given by

$$m = k(\text{volume}) = \frac{2}{3}k\pi r^3.$$

Using symmetry, we have

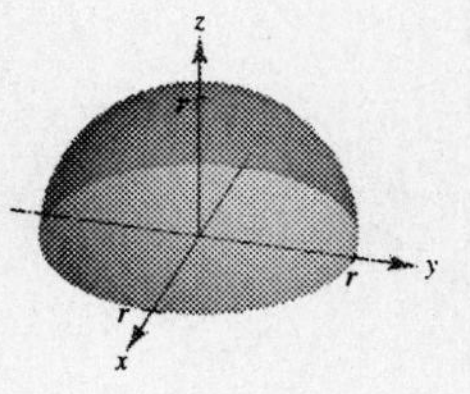


$$\begin{aligned}
M_{xy} &= \iiint_Q z(\text{density})\, dV \\
&= 4k\int_0^{\pi/2}\int_0^{\pi/2}\int_0^{r} \rho^3 \cos\phi \sin\, d\rho\, d\theta\, d\phi \\
&= \frac{1}{2}kr^4\int_0^{\pi/2}\int_0^{\pi/2} \sin 2\phi\, d\theta\, d\phi \\
&= \frac{1}{4}kr^4\pi\int_0^{\pi/2} \sin 2\phi\, d\phi \\
&= -\frac{1}{8}k\pi r^4\Big[\cos 2\phi\Big]_0^{\pi/2} = \frac{1}{4}k\pi r^4.
\end{aligned}$$

Therefore, $\bar{z} = \dfrac{M_{xy}}{m} = \dfrac{k\pi r^4/4}{2k\pi r^3/3} = \dfrac{3r}{8}$.

## Section 13.8 Change of Variables: Jacobians

**7.** Using the definition the Jacobian, we have

$$\frac{\partial(x, y)}{\partial(u, v)} = \begin{vmatrix} \dfrac{\partial x}{\partial u} & \dfrac{\partial y}{\partial u} \\ \dfrac{\partial x}{\partial v} & \dfrac{\partial y}{\partial v} \end{vmatrix} = \begin{vmatrix} e^u \sin v & e^u \cos v \\ e^u \cos c & -e^u \sin v \end{vmatrix} = -e^{2u}\sin^2 v - e^{2u}\cos^2 v = -e^{2u}.$$

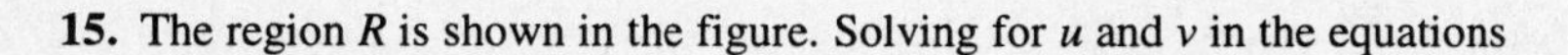

**15.** The region $R$ is shown in the figure. Solving for $u$ and $v$ in the equations

$$x = \sqrt{\frac{v}{u}} \quad \text{and} \quad y = \sqrt{uv}$$

produces $u = y/x$ and $v = xy$.

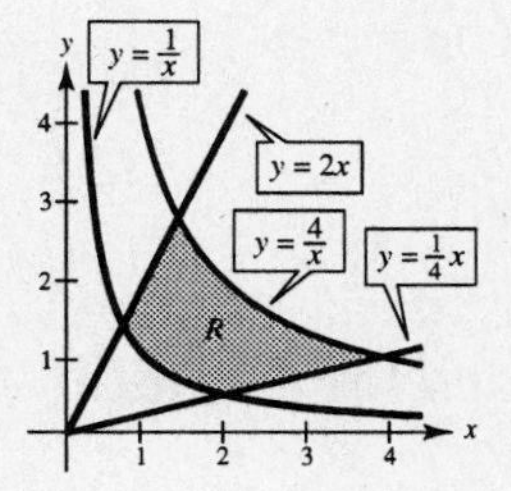


| *Bounds in the xy-plane* | | *Bounds in the uv-plane* |
|---|---|---|
| $\dfrac{y}{x} = \dfrac{1}{4}$ | $\rightarrow$ | $u = \dfrac{1}{4}$ |
| $\dfrac{y}{x} = 2$ | $\rightarrow$ | $u = 2$ |
| $xy = 1$ | $\rightarrow$ | $v = 1$ |
| $xy = 4$ | $\rightarrow$ | $v = 4$ |

The region $S$ is shown in the figure and the Jacobian for the transformation is

$$\frac{\partial(x, y)}{\partial(u, v)} = \begin{vmatrix} \dfrac{\partial x}{\partial u} & \dfrac{\partial y}{\partial u} \\ \dfrac{\partial x}{\partial v} & \dfrac{\partial y}{\partial v} \end{vmatrix} = \begin{vmatrix} -\dfrac{v^{1/2}}{2u^{3/2}} & \dfrac{v^{1/2}}{2u^{1/2}} \\ \dfrac{1}{2u^{1/2}v^{1/2}} & \dfrac{u^{1/2}}{2v^{1/2}} \end{vmatrix} = -\frac{1}{2u}.$$

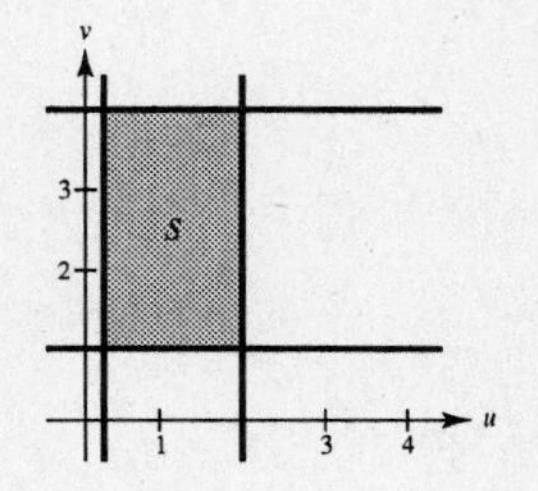


—CONTINUED—

**15. —CONTINUED—**

Therefore, we have

$$\begin{aligned}\int_R\int e^{-xy/2}\,dy\,dx &= \int_S\int e^{-v/2}\left|\frac{\partial(x,y)}{\partial(u,v)}\right|du\,dv\\
&= \frac{1}{2}\int_1^4\int_{1/4}^2 \frac{1}{u}e^{-v/2}\,du\,dv\\
&= \frac{1}{2}\int_1^4 e^{-v/2}\Big[\ln|u|\Big]_{1/4}^2\,dv\\
&= \frac{1}{2}\left(\ln 2 - \ln\frac{1}{4}\right)(-2)\int_1^4 e^{-v/2}\left(-\frac{1}{2}\right)dv\\
&= -\ln 8\Big[e^{-v/2}\Big]_1^4\\
&= -\ln 8(e^{-2} - e^{-1/2}) = \ln 8(e^{-1/2} - e^{-2}) \approx 0.9798.\end{aligned}$$

**19.** The region $R$ is bounded by the graphs of $x - y = 0$, $x - y = 5$, $x + 4y = 0$, and $x + 4y = 5$ (see figure). By letting $u = x - y$ and $v = x + 4y$, we have

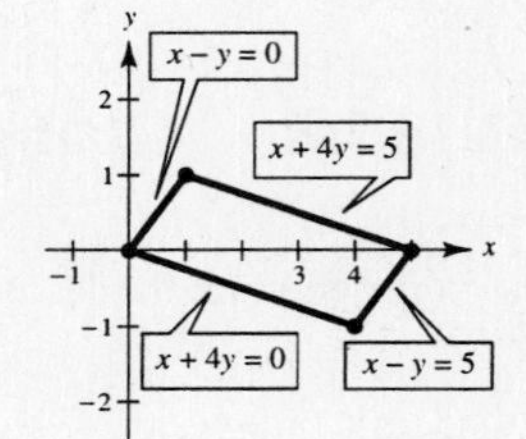


$$x = \frac{1}{5}(4u + v) \quad \text{and} \quad y = -\frac{1}{5}(u - v).$$

Thus, the Jacobian is

$$\frac{\partial(x,y)}{\partial(u,v)} = \begin{vmatrix}\dfrac{\partial x}{\partial u} & \dfrac{\partial y}{\partial u}\\[2mm] \dfrac{\partial x}{\partial v} & \dfrac{\partial y}{\partial v}\end{vmatrix} = \begin{vmatrix}\dfrac{4}{5} & -\dfrac{1}{5}\\[2mm] \dfrac{1}{5} & \dfrac{1}{5}\end{vmatrix} = \frac{1}{5}.$$

Therefore,

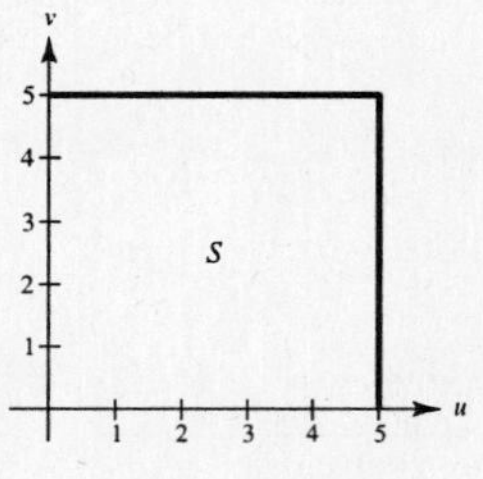

$$\begin{aligned}\int_R\int \sqrt{(x-y)(x+4y)}\,dy\,dx &= \int_S\int \sqrt{uv}\left|\frac{\partial(x,y)}{\partial(u,v)}\right|dv\,du\\
&= \frac{1}{5}\int_0^5\int_0^5 \sqrt{uv}\,dv\,du\\
&= \frac{2\sqrt{5}}{3}\int_0^5 \sqrt{u}\,du = \frac{100}{9}.\end{aligned}$$

## Review Exercises for Chapter 13

**7.** $\displaystyle\int_0^h\int_0^x \sqrt{x^2 + y^2}\,dy\,dx$

We choose to use polar coordinates since $r = \sqrt{x^2 + y^2}$. To rewrite the limits, we first find the equation of the line $x = h$ in polar coordinates.

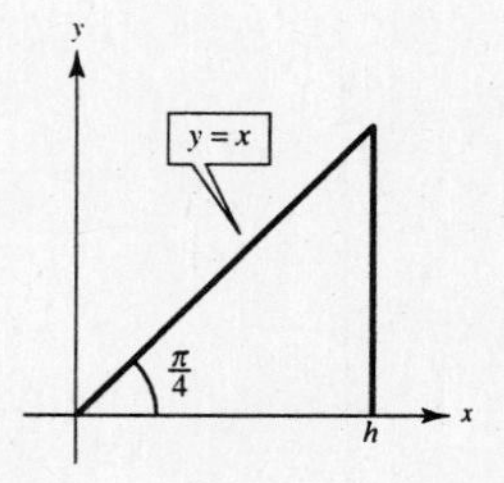


$$x = h$$

$$r\cos\theta = h$$

$$r = h\left(\frac{1}{\cos\theta}\right) = h\sec\theta$$

**—CONTINUED—**

**7. —CONTINUED—**

Therefore,

$$\int_0^h \int_0^x \sqrt{x^2 + y^2}\, dy\, dx = \int_0^{\pi/4} \int_0^{h \sec \theta} (r) r\, dr\, d\theta$$

$$= \frac{h^3}{3} \int_0^{\pi/4} \sec^3 \theta\, d\theta$$

$$= \frac{h^3}{3} \left[ \frac{\sec \theta \tan \theta}{2} + \frac{1}{2} \ln|\sec \theta + \tan \theta| \right]_0^{\pi/4}$$

$$= \frac{h^3}{6} \left[ \sqrt{2} + \ln\left(\sqrt{2} + 1\right) \right].$$

**19.** $$\iint_R f(x, y)\, dA = \int_{-5}^{3} \int_{-\sqrt{25-x^2}}^{\sqrt{25-x^2}} f(x, y)\, dy\, dx$$

$$= \int_{-5}^{-4} \int_{-\sqrt{25-y^2}}^{\sqrt{25-y^2}} f(x, y)\, dx\, dy + \int_{-4}^{4} \int_{-\sqrt{25-y^2}}^{3} f(x, y)\, dx\, dy + \int_{4}^{5} \int_{-\sqrt{25-y^2}}^{\sqrt{25-y^2}} f(x, y)\, dx\, dy$$

The area of $R$ is

$$A = 2 \int_{-5}^{3} \int_0^{\sqrt{25-x^2}} dy\, dx$$

$$= 2 \int_{-5}^{3} \sqrt{25 - x^2}\, dx$$

$$= 2\left(\frac{1}{2}\right) \left[ x\sqrt{25 - x^2} + 25 \arcsin\left(\frac{x}{5}\right) \right]_{-5}^{3}$$

$$= 3(4) + 25 \arcsin\left(\frac{3}{5}\right) - 0 - 25\left(-\frac{\pi}{2}\right)$$

$$= 12 + \frac{25\pi}{2} + 25 \arcsin\left(\frac{3}{5}\right) \approx 67.36.$$

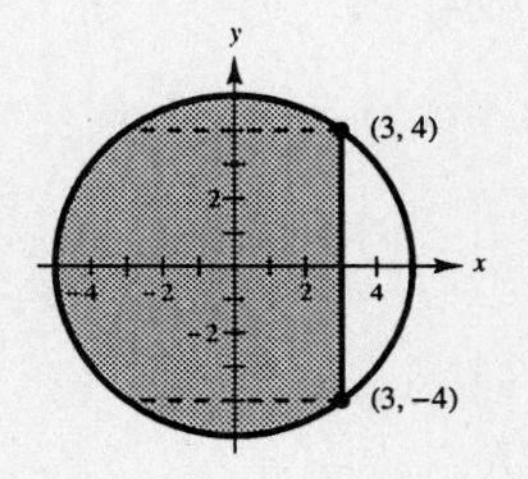


(**Note:** The area of the entire circle is $25\pi \approx 78.54$.)

**29.** The solid is outside the cylinder $x^2 + y^2 = 1$, inside the hyperboloid of one sheet $x^2 + y^2 - z^2 = 1$, above the $xy$-plane, and below the plane $z = h$ as shown in the figure. At a given height $z_0$ in cylindrical coordinates $r$ is bounded by 1 and

$$x^2 + y^2 - z_0^2 = 1$$

$$r^2 = 1 + z_0^2$$

$$r = \sqrt{1 + z_0^2}$$

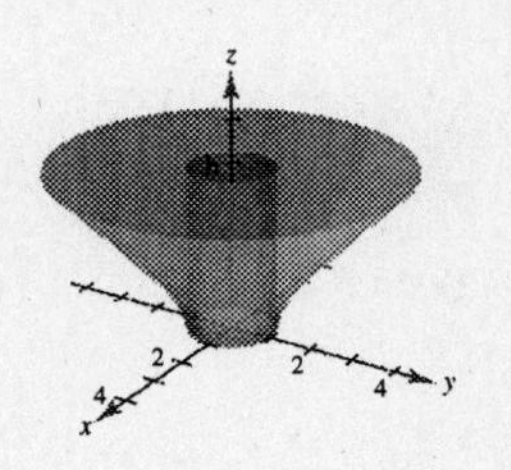


Therefore, by symmetry we have

$$V = 4 \int_0^h \int_0^{\pi/2} \int_1^{\sqrt{1+z^2}} r\, dr\, d\theta\, dz$$

$$= 2 \int_0^h \int_0^{\pi/2} (1 + z^2 - 1)\, d\theta\, dz$$

$$= \pi \int_0^h z^2\, dz = \left[ \pi\left(\frac{1}{3} z^3\right) \right]_0^h = \frac{\pi h^3}{3}.$$

**35.** $f(x, y) = \begin{cases} kxye^{-(x+y)}, & x \geq 0, y \geq 0 \\ 0, & \text{elsewhere} \end{cases}$

If $f$ is a joint density function, then $f(x, y) \geq 0$ for all $(x, y)$, and

$$\int_{-\infty}^{\infty}\int_{-\infty}^{\infty} f(x, y)\, dA = 1.$$

Use Integration by Parts to evaluate the improper integral.

$$\int_{-\infty}^{\infty}\int_{-\infty}^{\infty} f(x, y)\, dA = \int_0^{\infty}\int_0^{\infty} kxye^{-(x+y)}\, dy\, dx$$

$$= \int_0^{\infty}\left[-kxe^{-(x+y)}(y + 1)\right]_0^{\infty} dx$$

$$= \int_0^{\infty} kxe^{-x}\, dx$$

$$= \left[-k(x + 1)e^{-x}\right]_0^{\infty} = k$$

Therefore, $k = 1$. Use the same integration steps shown above to find the required probability.

$$P(0 \leq x \leq 1, 0 \leq y \leq 1) = \int_0^1\int_0^1 kxye^{-(x+y)}\, dy\, dx$$

$$\approx 0.070$$

**41.** $f(x, y) = 16 - x^2 - y^2,\ R\{(x, y): x^2 + y^2 \leq 16\}$

Since $z = 16 - x^2 - y^2$,

$$\frac{\partial z}{\partial x} = -2x \quad \text{and} \quad \frac{\partial z}{\partial y} = -2y.$$

$$S = \int\!\!\int_R \sqrt{1 + \left(\frac{\partial z}{\partial x}\right)^2 + \left(\frac{\partial z}{\partial y}\right)^2}\, dy\, dx$$

$$= \int_{-4}^{4}\int_{-\sqrt{16-x^2}}^{\sqrt{16-x^2}} \sqrt{1 + 4x^2 + 4y^2}\, dy\, dx$$

$$= 4\int_0^4\int_0^{\sqrt{16-x^2}} \sqrt{1 + 4(x^2 + y^2)}\, dy\, dx$$

$$= \frac{1}{2}\int_0^{\pi/2}\int_0^4 \sqrt{1 + 4r^2}(8r)\, dr\, d\theta$$

$$= \frac{1}{3}\int_0^{\pi/2} (65^{3/2} - 1)\, d\theta = \frac{\pi}{6}(65\sqrt{65} - 1)$$

**47.** $x^2 + y^2 + z^2 = a^2$ (first octant)

Because of its symmetry, the coordinates of the center of mass are equal. Let $k$ be the constant density; then the mass is

$$m = k(\text{volume}) = k\left(\frac{1}{8}\right)\left(\frac{4}{3}\pi a^3\right) = \frac{k}{6}\pi a^3.$$

Now,

$$M_{xy} = \int\!\!\int\!\!\int_S z(\text{density})\, dS \qquad \text{(rectangular coordinates)}$$

$$= k\int_0^{\pi/2}\int_0^{\pi/2}\int_0^a (\rho\cos\phi)(\rho^2\sin\phi)\, d\rho\, d\theta\, d\phi \qquad \text{(spherical coordinates)}$$

$$= \frac{ka^4}{4}\int_0^{\pi/2}\int_0^{\pi/2} \cos\phi\sin\phi\, d\theta\, d\phi$$

$$= \frac{k\pi a^4}{8}\int_0^{\pi/2} \cos\phi\sin\phi\, d\phi$$

$$= \frac{k\pi a^4}{8}\left[\frac{1}{2}\sin^2\phi\right]_0^{\pi/2} = \frac{k\pi a^4}{16}.$$

Therefore,

$$\bar{x} = \bar{y} = \bar{z} = \frac{M_{xy}}{m} = \frac{k\pi a^4/16}{k\pi a^3/6} = \frac{3a}{8}.$$

# CHAPTER 14
## Vector Analysis

# CHAPTER 14
# Vector Analysis

## Section 14.1 Vector Fields

**Solutions to Selected Odd-Numbered Exercises**

**9.** $\mathbf{F}(x, y) = x\mathbf{i} + y\mathbf{i}$

We will plot vectors of equal magnitude and, in this case, they lie along circles given by

$$\|\mathbf{F}(x, y)\| = \sqrt{x^2 + y^2} = c \Rightarrow x^2 + y^2 = c^2$$

For $c = 1$, sketch several vectors $x\mathbf{i} + y\mathbf{j}$ of magnitude 1 on the circle given $x^2 + y^2 = 1$.
For $c = 4$, sketch several vectors $x\mathbf{i} + y\mathbf{j}$ of magnitude 2 on the circle given by $x^2 + y^2 = 4$.
(see figure)

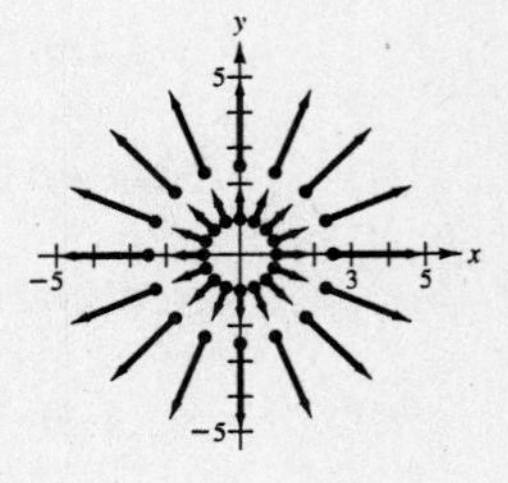


**23.** $f(x, y, z) = a - ye^{x^2}$

The gradient vector field for $f$ is

$$\mathbf{F}(x, y, z) = f_x(x, y, y)\mathbf{i} + f_y(x, y, z)\mathbf{j} + f_z(x, y, z)\mathbf{k}$$
$$= -2xye^{x^2}\mathbf{i} - e^{x^2}\mathbf{j} + \mathbf{k}.$$

**29.** Since $\mathbf{F}(x, y) = 2xye^{x^2y}\mathbf{i} + x^2e^{x^2y}\mathbf{j}$, it follows that $\mathbf{F}$ is conservative since

$$\frac{\partial}{\partial y}[2xye^{x^2y}] = 2x^3ye^{x^2y} + 2xe^{x^2y} = \frac{\partial}{\partial x}[x^2e^{x^2y}].$$

If $f$ is a function such that $\nabla f(x, y) = f_x(x, y)\mathbf{i} + f_y(x, y)\mathbf{j}$, then we have

$$f_x(x, y) = 2xye^{x^2y} \quad \text{and} \quad f_y(x, y) = x^2e^{x^2y}.$$

To reconstruct the function $f$ from these two partial derivatives, integrate $f_x(x, y)$ with respect to $x$ and $f_y(x, y)$ with respect to $y$ as follows.

$$f(x, y) = \int f_x(x, y)\,dx = \int 2xye^{x^2y}\,dx = e^{x^2y} + g(y) + K$$

$$f(x, y) = \int f_y(x, y)\,dy = \int x^2e^{x^2y}\,dy = e^{x^2y} + h(x) + K$$

These two expressions for $f(x, y)$ are the same if $g(y) = h(x) = 0$. Therefore, we have

$$f(x, y) = e^{x^2y} + K.$$

**35.** $\mathbf{F}(x, y, z) = xyz\mathbf{i} + y\mathbf{j} + z\mathbf{k}$

It follows from the definition of the curl that

$$\textbf{curl}\,\mathbf{F}(x, y, z) = \begin{vmatrix} \mathbf{i} & \mathbf{j} & \mathbf{k} \\ \frac{\partial}{\partial x} & \frac{\partial}{\partial y} & \frac{\partial}{\partial z} \\ xyz & y & z \end{vmatrix} = \begin{vmatrix} \frac{\partial}{\partial y} & \frac{\partial}{\partial z} \\ y & z \end{vmatrix}\mathbf{i} - \begin{vmatrix} \frac{\partial}{\partial x} & \frac{\partial}{\partial z} \\ xyz & z \end{vmatrix}\mathbf{j} + \begin{vmatrix} \frac{\partial}{\partial x} & \frac{\partial}{\partial y} \\ xyz & y \end{vmatrix}\mathbf{k}$$

$$= (0 - 0)\mathbf{i} - (0 - xy)\mathbf{j} + (0 - xz)\mathbf{k} = xy\mathbf{j} - xz\mathbf{k}.$$

Therefore, $\textbf{curl}\,\mathbf{F}(1, 2, 1) = 2\mathbf{j} - \mathbf{k}$.

**39.** $\mathbf{F}(x, y, z) = \left(\arctan \frac{x}{y}\right)\mathbf{i} + \left(\ln\sqrt{x^2 + y^2}\right)\mathbf{j} + \mathbf{k}$

$$\mathbf{curl\ F}(x, y, z) = \begin{vmatrix} \mathbf{i} & \mathbf{j} & \mathbf{k} \\ \dfrac{\partial}{\partial x} & \dfrac{\partial}{\partial y} & \dfrac{\partial}{\partial z} \\ \arctan \dfrac{x}{y} & \ln\sqrt{x^2 + y^2} & 1 \end{vmatrix} = (0 - 0)\mathbf{i} - (0 - 0)\mathbf{j} + \left[\frac{x}{x^2 + y^2} - \frac{-x/y^2}{1 + (x^2/y^2)}\right]\mathbf{k} = \frac{2x}{x^2 + y^2}\mathbf{k}$$

**43.** $\mathbf{F}(x, y, z) = \sin y\mathbf{i} - x\cos y\mathbf{j} + \mathbf{k} = M\mathbf{i} + N\mathbf{j} + P\mathbf{k}$

Since

$$\frac{\partial}{\partial y} = 0 = \frac{\partial N}{\partial z},\ \frac{\partial P}{\partial x} = 0 = \frac{\partial M}{\partial z},\ \frac{\partial N}{\partial x} = -\cos y \neq \cos y = \frac{\partial M}{\partial y},$$

it follows that $\mathbf{F}$ is not conservative.

**47.** $\mathbf{F}(x, y, z) = \frac{1}{y}\mathbf{i} - \frac{x}{y^2}\mathbf{j} + (2x - 1)\mathbf{k} = M\mathbf{i} + N\mathbf{j} + P\mathbf{k}$

Since

$$\frac{\partial P}{\partial y} = 0 = \frac{\partial N}{\partial z},\ \frac{\partial P}{\partial x} = 0 = \frac{\partial M}{\partial z},\ \frac{\partial N}{\partial x} = -\frac{1}{y^2} = \frac{\partial M}{\partial y},$$

it follows that $\mathbf{F}$ is conservative. Now, if $f$ is a function such that $\mathbf{F}(x, y, z) = \nabla f(x, y, z)$, then

$$f_x(x, y, z) = \frac{1}{y},\ f_y(x, y, z) = -\frac{x}{y^2},\ f_z(x, y, z) = 2z - 1,$$

and by integrating with respect to $x$, $y$, and $z$ respectively, we obtain

$$f(x, y, z) = \int M\,dx = \int \frac{1}{y}\,dx = \frac{x}{y} + g(y, z) + K$$

$$f(x, y, z) = \int N\,dy = \int -\frac{x}{y^2}\,dy = \frac{x}{y} + h(x, z) + K$$

$$f(x, y, z) = \int P\,dz = \int (2z - 1)\,dz = z^2 - z + k(x, y) + K.$$

By comparing these three versions of $f$, we can conclude that

$$g(y, z) = z^2 - z,\ h(x, z) = z^2 - z, \text{ and } k(x, y) = \frac{x}{y}.$$

Therefore,

$$f(x, y, z) = \frac{x}{y} + z^2 - z + K.$$

**51.** $\mathbf{F}(x, y, z) = xyz\mathbf{i} + y\mathbf{j} + z\mathbf{k}$

From Exercise 35 we have $\mathbf{curl\ F}(x, y, z) = xy\mathbf{j} - xz\mathbf{k}$. Thus,

$$\mathbf{curl}[(\mathbf{curl\ F})(x, y, z)] = \begin{vmatrix} \mathbf{i} & \mathbf{j} & \mathbf{k} \\ \dfrac{\partial}{\partial x} & \dfrac{\partial}{\partial y} & \dfrac{\partial}{\partial z} \\ 0 & xy & -xz \end{vmatrix} = (0 - 0)\mathbf{i} - (-z - 0)\mathbf{j} + (y - 0)\mathbf{k} = z\mathbf{j} + y\mathbf{k}.$$

**55.** $\mathbf{F}(x, y, z) = \sin x\mathbf{i} + \cos y\mathbf{j} + z^2\mathbf{k}$

By definition,

$$\text{div } \mathbf{F}(x, y, z) = \frac{\partial}{\partial x}[\sin x] + \frac{\partial}{\partial y}[\cos y] + \frac{\partial}{\partial z}[z^2]$$

$$= \cos x - \sin y + 2z.$$

**63.** $\mathbf{F}(x, y, z) = xyz\mathbf{i} + y\mathbf{j} + z\mathbf{k}$

From Exercise 35 we have $\mathbf{curl\ F}(x, y, z) = xy\mathbf{j} - xz\mathbf{k}$. By definition,

$$\text{div}(\mathbf{curl\ F}) = \frac{\partial}{\partial x}[0] + \frac{\partial}{\partial y}[xy] + \frac{\partial}{\partial z}[-xz]$$

$$= x - x = 0.$$

## Section 14.2 Line Integrals

**9.** $\int_C (x^2 + y^2 + z^2)\, ds$

$C$: $\mathbf{r}(t) = \sin t\mathbf{i} + \cos t\mathbf{j} + 8t\mathbf{k},\ 0 \le t \le \dfrac{\pi}{2}$

Since $r'(t) = \cos t\mathbf{i} - \sin t\mathbf{j} + 8\mathbf{k}$, we have

$$\begin{aligned} ds = \|\mathbf{r}'(t)\|\, dt &= \sqrt{[x'(t)]^2 + [y'(t)]^2 + [z'(t)]^2}\, dt \\ &= \sqrt{\cos^2 t + (-\sin t)^2 + 8^2}\, dt = \sqrt{65}\, dt. \end{aligned}$$

It follows that

$$\begin{aligned} \int_C (x^2 + y^2 + z^2)\, ds &= \int_0^{\pi/2} (\sin^2 t + \cos^2 t + 64t^2)\sqrt{65}\, dt \\ &= \sqrt{65}\int_0^{\pi/2} (1 + 64t^2)\, dt = \sqrt{65}\left[t + \frac{64}{3}t^3\right]_0^{\pi/2} \\ &= \frac{\sqrt{65}\pi}{6}(3 + 16\pi^2). \end{aligned}$$

**13.** $\int_C (x^2 + y^2)\, ds$

$C$: $x^2 + y^2 = 1$ from $(1, 0)$ counterclockwise to $(0, 1)$.

Since the path is one-fourth the unit circle, it can be represented by $\mathbf{r}(t) = \cos t\mathbf{i} + \sin t\mathbf{j}$ for $0 \le t \le \pi/2$. Therefore,

$$\begin{aligned} \mathbf{r}'(t) &= -\sin t\mathbf{i} + \cos t\mathbf{j} \\ ds = \|\mathbf{r}'(t)\|\, dt &= \sqrt{(-\sin t)^2 + (\cos t)^2}\, dt = dt. \end{aligned}$$

It follows that

$$\int_C (x^2 + y^2)\, ds = \int_0^{\pi/2} (\cos^2 t + \sin^2 t)\, dt = \int_0^{\pi/2} dt = \frac{\pi}{2}.$$

**17.** $\int_C \left(x + 4\sqrt{y}\right) ds$

$C$: counterclockwise around the triangle with vertices $(0, 0)$, $(1, 0)$, and $(0, 1)$.

Path $C$ has parts as shown in the figure.

$$\begin{aligned} &C_1: x = t, && y = 0, && 0 \le t \le 1, && ds = \sqrt{1 + 0}\, dt = dt \\ &C_2: x = 1 - t, && y = t, && 0 \le t \le 1, && ds = \sqrt{1 + 1}\, dt = \sqrt{2}\, dt \\ &C_3: x = 0, && y = 1 - t, && 0 \le t \le 1, && ds = \sqrt{0 + 1}\, dt = dt \end{aligned}$$

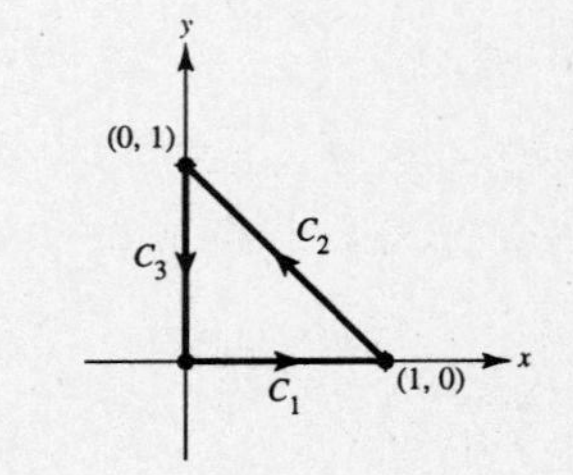


Therefore,

$$\begin{aligned} \int_C \left(x + 4\sqrt{y}\right) dx &= \int_0^1 t\, dt + \int_0^1 \left[(1 - t) + 4\sqrt{t}\right]\sqrt{2}\, dt + \int_0^1 4\sqrt{1 - t}\, dt \\ &= \left[\frac{t^2}{2}\right]_0^1 + \sqrt{2}\left[t - \frac{t^2}{2} + \frac{8}{3}t^{3/2}\right]_0^1 - \left[\frac{8}{3}(1 - t)^{3/2}\right]_0^1 \\ &= \frac{1}{2} + \sqrt{2}\left(\frac{19}{6}\right) + \frac{8}{3} = \frac{19}{6}(1 + \sqrt{2}). \end{aligned}$$

**23.** $\mathbf{F}(x, y) = 3x\mathbf{i} + 4y\mathbf{j}$

$C$: $\mathbf{r}(t) = (2\cos t)\mathbf{i} + (2\sin t)\mathbf{j},\ 0 \le t \le \dfrac{\pi}{2}$

Since $x(t) = 2\cos t$ and $y(t) = 2\sin t$, the vector field can be written $\mathbf{F}(x, y) = 6\cos t\mathbf{i} + 8\sin t\mathbf{j}$. Use the fact that $\mathbf{r}'(t) = -2\sin t\mathbf{i} + 2\cos t\mathbf{j}$ and write the following.

$$\int_C \mathbf{F}\cdot d\mathbf{r} = \int_a^b \mathbf{F}(x(t), y(t))\cdot \mathbf{r}'(t)\,dt$$

$$= \int_0^{\pi/2} (-12\sin t\cos t + 16\sin t\cos t)\,dt$$

$$= 4\int_0^{\pi/2} \sin t\cos t\,dt = 4\left[\frac{\sin^2 t}{2}\right]_0^{\pi/2} = 2$$

**29.** $\mathbf{F}(x, y) = -x\mathbf{i} - 2y\mathbf{j}$

$C$: $y = x^3$ from $(0, 0)$ to $(2, 8)$

Since the path of $C$ is given by $x = t$ and $y = t^3$, with $0 \le t \le 2$, we have

$$\mathbf{F}(t) = -t\mathbf{i} - 2t^3\mathbf{j} \quad \text{and} \quad \mathbf{r}'(t) = \mathbf{i} + 3t^2\mathbf{j}.$$

Therefore,

$$W = \int_C \mathbf{F}(t)\cdot\mathbf{r}'(t)\,dt = \int_0^2 (-t - 6t^5)\,dt$$

$$= \left[-\frac{t^2}{2} - t^6\right]_0^2 = -66.$$

**39.** $\mathbf{F}(x, y) = (x^3 - 2x^2)\mathbf{i} + \left(x - \dfrac{y}{2}\right)\mathbf{j}$

$C$: $\mathbf{r}(t) = t\mathbf{i} + t^2\mathbf{j}$

Since the path of $C$ is given by $x = t$ and $y = t^2$, we have $\mathbf{F}(x, y) = (t^3 - 2t^2)\mathbf{i} + \left(t - \dfrac{t^2}{2}\right)\mathbf{j}$ and $\mathbf{r}'(t) = \mathbf{i} + 2t\mathbf{j}$.

Therefore,

$$\int_C \mathbf{F}\cdot d\mathbf{r} = \int_a^b \mathbf{F}(x(t), y(t))\cdot\mathbf{r}'(t)\,dt$$

$$= \int_a^b (t^3 - 2t^2 + 2t^2 - t^3)\,dt = \int_a^b 0\,dt = 0.$$

**51.** $\displaystyle\int_C (2x - y)\,dx + (x + 3y)\,dy$

$C$: $x = t$ and $y = 2t^2$ from $(0, 0)$ to $(2, 8)$

The path $C$ is given by $x = t$, $y = 2t^2$, with $0 \le t \le 2$, $dx = dt$, and $dy = 4t\,dt$. Therefore,

$$\int_C (2x - y)\,dx + (x + 3y)\,dy = \int_0^2 (2t - 2t^2)\,dt + (t + 6t^2)4t\,dt$$

$$= \int_0^2 (2t + 2t^2 + 24t^3)\,dt$$

$$= \left[t^2 + \frac{2t^3}{3} + 6t^4\right]_0^2$$

$$= 4 + \frac{16}{3} + 96 = \frac{316}{3}.$$

**55.** $f(x, y) = xy$

$C$: $x^2 + y^2 = 1$ from $(1, 0)$ to $(0, 1)$

We represent $C$ parametrically as $x = \cos t$ and $y = \sin t$ for $0 \le t \le \pi/2$. Then

$$f(x, y) = xy = \cos t\sin t \quad \text{and} \quad ds = \sqrt{[x'(t)]^2 + [y'(t)]^2}\,dt = \sqrt{(-\sin t)^2 + (\cos t)^2}\,dt = dt.$$

Therefore,

$$\text{area} = \int_C f(x, y)\,ds$$

$$= \int_0^{\pi/2} \cos t\sin t\,dt = \left[\frac{1}{2}\sin^2 t\right]_0^{\pi/2} = \frac{1}{2}.$$

**61.** (a)
$$z = f(x, y) = 1 + y^2$$
$$\mathbf{r}(t) = 2\cos t\mathbf{i} + 2\sin t\mathbf{j}$$
$$\mathbf{r}'(t) = -2\sin t\mathbf{i} + 2\cos t\mathbf{j}$$
$$\|\mathbf{r}'(t)\| = \sqrt{(-2\sin t)^2 + (2\cos t)^2} = 2$$
$$S = \int_C f(x, y)\,ds$$
$$= \int_a^b f(x(t), y(t))\|\mathbf{r}'(t)\|\,dt$$
$$= \int_0^{2\pi} (1 + 4\sin^2 t)(2)\,dt$$
$$= \int_0^{2\pi} (6 - 4\cos 2t)\,dt$$
$$= \Big[6t - 2\sin 2t\Big]_0^{2\pi} = 12\pi \approx 37.70 \text{ sq cm}$$

(b) The volume of steel is approximated by the product of the thickness of the steel and the surface area of the component.

$$0.2(12\pi) = \frac{12\pi}{5} \approx 7.54 \text{ cu cm}$$

(c) A sketch of the component is shown in the figure.

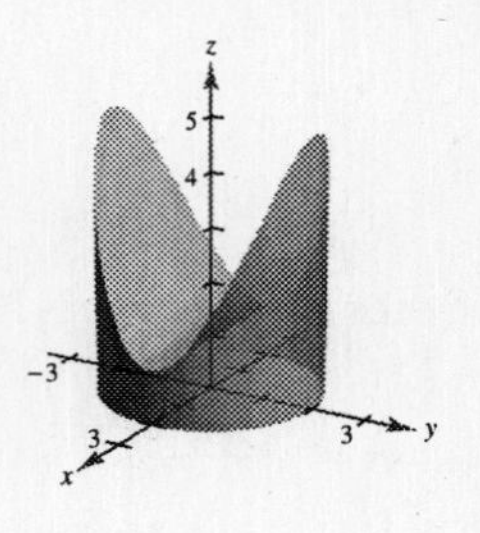


## Section 14.3 Conservative Vector Fields and Independence of Path

**3.** $\mathbf{F}(x, y) = y\mathbf{i} - x\mathbf{j}$

(a) For $\mathbf{r}_1(\theta) = \sec\theta\mathbf{i} + \tan\theta\mathbf{j}$, we have $\mathbf{F}(x, y) = \tan\theta\mathbf{i} - \sec\theta\mathbf{j}$ and $\mathbf{r}'(\theta) = \sec\theta\tan\theta\mathbf{i} + \sec^2\theta\mathbf{j}$. Therefore,

$$\int_C \mathbf{F}\cdot d\mathbf{r}_1 = \int_0^{\pi/3} \mathbf{F}\cdot\mathbf{r}_1'\,d\theta = \int_0^{\pi/3} (\sec\theta\tan^2\theta - \sec^3\theta)\,d\theta$$
$$= -\int_0^{\pi/3} \sec\theta\,d\theta = \Big[-\ln|\sec\theta + \tan\theta|\Big]_0^{\pi/3} = -\ln(2 + \sqrt{3})$$

(b) For $\mathbf{r}_2(t) = \sqrt{t+1}\,\mathbf{i} + \sqrt{t}\,\mathbf{j}$, we have $\mathbf{F}(t)\sqrt{t}\mathbf{i} - \sqrt{t+1}\,\mathbf{j}$ and $\mathbf{r}_2'(t) = \dfrac{1}{2\sqrt{t+1}}\mathbf{i} + \dfrac{1}{2\sqrt{t}}\mathbf{j}$. Therefore,

$$\int_C \mathbf{F}\cdot d\mathbf{r}_2 = \int_0^3 \mathbf{F}\cdot\mathbf{r}_2'\,dt$$
$$= \int_0^3 \left(\frac{\sqrt{t}}{2\sqrt{t+1}} - \frac{\sqrt{t+1}}{2\sqrt{t}}\right)dt = \frac{1}{2}\int_0^3 \frac{-1}{\sqrt{t+1}\sqrt{t}}\,dt.$$

If $u = \sqrt{t+1}$, then $u^2 = t + 1$, $u^2 - 1 = t$, $2u\,du = dt$, and $\sqrt{u^2 - 1} = \sqrt{t}$. Thus,

$$\frac{1}{2}\int_0^3 \frac{-1}{\sqrt{t+1}\sqrt{t}}\,dt = -\int_1^2 \frac{du}{\sqrt{u^2 - 1}} = \Big[-\ln|u + \sqrt{u^2 + 1}|\Big]_1^2 = -\ln(2 + \sqrt{3}).$$

**9.** The vector field $\mathbf{F}(x, y, z) = M\mathbf{i} + N\mathbf{j} + P\mathbf{k}$ is conservative if and only if

$$\frac{\partial P}{\partial y} = \frac{\partial N}{\partial z},\ \frac{\partial P}{\partial x} = \frac{\partial M}{\partial z},\ \text{and}\ \frac{\partial N}{\partial x} = \frac{\partial M}{\partial y}.$$

For the vector field $\mathbf{F}(x, y, z) = y^2z\mathbf{i} + 2xyz\mathbf{j} + xy^2\mathbf{k}$, we have

$$\frac{\partial P}{\partial y} = 2xy = \frac{\partial N}{\partial z}$$
$$\frac{\partial P}{\partial x} = y^2 = \frac{\partial M}{\partial z}$$
$$\frac{\partial N}{\partial x} = 2yz = \frac{\partial M}{\partial y}.$$

Therefore, $\mathbf{F}$ is conservative. A second method for determining whether $\mathbf{F}$ is conservative is to find **curl** $\mathbf{F}(x, y, z)$. $\mathbf{F}$ is conservative if and only if **curl** $\mathbf{F}(x, y, z) = \mathbf{0}$.

**17.** $\int_C 2xy\,dx + (x^2 + y^2)\,dy$

(a) $C$: ellipse $\left(\frac{x^2}{25}\right) + \left(\frac{y^2}{16}\right) = 1$ from $(5, 0)$ to $(0, 4)$

Observe that the vector field $\mathbf{F}(x, y) = 2xy\mathbf{i} + (x^2 + y^2)\mathbf{j}$ is conservative, since

$$\frac{\partial}{\partial y}[2xy] = 2x = \frac{\partial}{\partial x}[x^2 + y^2].$$

Therefore, the line integral is independent of the path and we can replace the path along the ellipse from $(5, 0)$ to $(0, 4)$ with a path which will simplify the integration. One possibility is the path along the coordinate axes from $(5, 0)$ to $(0, 0)$ and then from $(0, 0)$ to $(0, 4)$. Along the path from $(5, 0)$ to $(0, 0)$ we have $y = 0$ and $dy = 0$. On the path from $(0, 0)$ to $(0, 4)$ we have $x = 0$ and $dx = 0$. Hence,

$$\int_C 2xy\,dx + (x^2 + y^2)\,dy = \int_5^0 0\,dx + (x^2)0 + \int_0^4 (0)(0) + (0 + y^2)\,dy$$
$$= \left[\frac{y^3}{3}\right]_0^4 = \frac{64}{3}.$$

(b) $C$: parabola $y = 4 - x^2$ from $(2, 0)$ to $(0, 4)$

Using the same method as in part (a) replace the path along the parabola by the path along the axes from $(2, 0)$ to $(0, 0)$ and then from $(0, 0)$ to $(0, 4)$. Thus,

$$\int_C 2xy\,dx + (x^2 + y^2)\,dy = \int_2^0 0\,dx + (x^2 + 0)(0) + \int_0^4 (0)(0) + (0 + y^2)\,dy$$
$$= \left[\frac{y^3}{3}\right]_0^4 = \frac{64}{3}.$$

(Since the line integral is path-independent, we could have used the Fundamental Theorem.)

**21.** $\mathbf{F}(x, y, z) = (2y + x)\mathbf{i} + (x^2 - z)\mathbf{j} + (2y - 4z)\mathbf{k}$

(a) $\mathbf{r}_1(t) = t\mathbf{i} + t^2\mathbf{j} + \mathbf{k},\ 0 \le t \le 1$

Along the path of $\mathbf{r}_1(t)$, we have

$$\mathbf{F}(x, y, z) = (2t^2 + t)\mathbf{i} + (t^2 - 1)\mathbf{j} + (2t^2 - 4)\mathbf{k} \quad \text{and} \quad \mathbf{r}_1{}' = \mathbf{i} + 2t\mathbf{j}.$$

Therefore,

$$\int_C \mathbf{F} \cdot d\mathbf{r}_1 = \int_a^b \mathbf{F}(x(t), y(t), z(t)) \cdot r_1{}'(t)\,dt$$
$$= \int_0^1 (2t^3 + 2t^2 - t)\,dt = \frac{2}{3}.$$

(b) $\mathbf{r}_2(t) = t\mathbf{i} + t\mathbf{j} + (2t - 1)^2\mathbf{k}$

Along the path of $\mathbf{r}_2(t)$, we have

$$\mathbf{F}(x, y, z) = (2t + t)\mathbf{i} + [t^2 - (2t - 1)^2]\mathbf{j} + [2t - 4(2t - 1)^2]\mathbf{k}$$
$$= 3t\mathbf{i} + (-3t^2 + 4t - 1)\mathbf{j} + (-16t^2 + 18t - 4)\mathbf{k}$$

and

$$\mathbf{r}_2{}'(t) = \mathbf{i} + \mathbf{j} + 4(2t - 1)\mathbf{k}.$$

Therefore,

$$\int_C \mathbf{F} \cdot d\mathbf{r}_2 = \int_a^b \mathbf{F}(x(t), y(t), z(t)) \cdot \mathbf{r}_2{}'(t)\,dt$$
$$= \int_0^1 (-128t^3 + 205t^2 - 97t + 15)\,dt = \frac{17}{6}.$$

**29.** $\displaystyle\int_C e^x \sin y\,dx + e^x \cos y\,dy$

$C$: $x = \theta - \sin\theta$, $y = 1 - \cos\theta$ from $(0, 0)$ to $(2\pi, 0)$

Since

$$\frac{\partial}{\partial y}[e^x \sin y] = e^x \cos y = \frac{\partial}{\partial x}[e^x \cos y],$$

the integral is path-independent. Therefore, we can evaluate the line integral by using the Fundamental Theorem. We begin by finding the potential function for the vector field $\mathbf{F}(x, y) = e^x \sin y\mathbf{i} + e^x \cos y\mathbf{j}$. If $f$ is a potential function of $\mathbf{F}$, then

$$f_x(x, y) = e^x \sin y \quad \text{and} \quad f_y(x, y) = e^x \cos y$$

and we have

$$f(x, y) = \int f_x(x, y)\,dx = \int e^x \sin y\,dx = e^x \sin y + g(y) + K$$

$$f(x, y) = \int f_y(x, y)\,dy = \int e^x \cos y\,dy = e^x \sin y + h(x) + K$$

These two expressions for $f(x, y)$ are the same if $g(y) = h(x) = 0$. Therefore, we have

$$f(x, y) = e^x \sin y + K$$

and

$$\int_C e^x \sin y\,dx + e^x \cos y\,dy = f(2\pi, 0) - f(0, 0) = e^{2\pi}(0) - e^0(0) = 0.$$

[Since this line integral is path-independent, we could have integrated along the $x$-axis from $(0, 0)$ to $(2\pi, 0)$ with $y = 0$ and $dy = 0$. This would have given the result of zero immediately.]

**35.** $\mathbf{F}(x, y) = 9x^2y^2\mathbf{i} + (6x^3y - 1)\mathbf{j}$

$\mathbf{F}$ is conservative because

$$\frac{\partial}{\partial y}[9x^2y^2] = 18x^2y \quad \text{and} \quad \frac{\partial}{\partial x}[6x^3y - 1] = 18x^2y.$$

If $f$ is a function such that $\nabla f = \mathbf{F}$, then

$$f_x(x, y) = 9x^2y^2 \quad \text{and} \quad f_y(x, y) = 6x^3y - 1.$$

$$f(x, y) = \int f_x(x, y)\,dx = \int 9x^2y^2\,dx = 3x^3y + g(y)$$

$$f(x, y) = \int f_y(x, y)\,dy = \int (6x^3y - 1)\,dx = 3x^3y^2 - y + h(x)$$

It follows that

$$g(y) = -y + K,\ h(x) = K, \text{ and } f(x, y) = 3x^3y^2 - y + K.$$

Therefore, the work done by $\mathbf{F}$ in moving an object from $P(0, 0)$ to $Q(5, 9)$ along any path $C$ is

$$W = \int_C \mathbf{F} \cdot d\mathbf{r} = f(5, 9) - f(0, 0) = 30{,}366.$$

## Section 14.4 Green's Theorem

**3.** $\int_C y^2\,dx + x^2\,dy$

$C$: boundary of the region lying between $y = x$ and $x = \frac{x^2}{4}$

(a) As a line integral define $C_1$ and $C_2$ as shown in the figure.

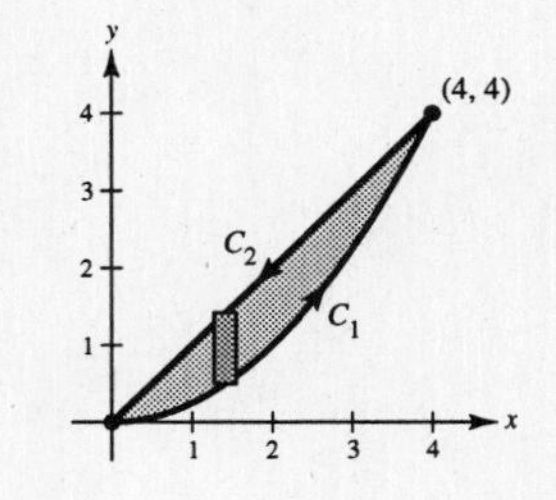


$$C_1\colon\ x = t,\ y = \frac{t^2}{4},\ dx = dt,\ dy = \frac{t}{2}\,dt,\ 0 \le t \le 4$$

$$C_2\colon\ x = 4 - t,\ y = 4 - t,\ dx = dy = -dt,\ 0 \le t \le 4$$

Thus,

$$\begin{aligned}\int_C y^2\,dx + x^2\,dy &= \int_{C_1}\left(\frac{t^4}{16} + \frac{t^3}{2}\right)dt + \int_{C_2} 2(4-t)^2(-dt)\\ &= \int_0^4\left(\frac{t^4}{16} + \frac{t^3}{2}\right)dt + 2\int_0^4 (4-t)^2(-1)\,dt\\ &= \left[\frac{t^5}{5(16)} + \frac{t^4}{2(4)} + \frac{2(4-t)^3}{3}\right]_0^4\\ &= \frac{64}{5} + 32 - \frac{128}{3} = \frac{32}{15}.\end{aligned}$$

(b) By Green's Theorem we have

$$\begin{aligned}\int_R\!\!\int\left(\frac{\partial N}{\partial x} - \frac{\partial M}{\partial y}\right)dA &= \int_R\!\!\int (2x - 2y)\,dy\,dx\\ &= \int_0^4\!\int_{x^2/4}^{x} (2x - 2y)\,dy\,dx = \int_0^4\left[2xy - y^2\right]_{x^2/4}^{x} dx\\ &= \int_0^4\left(x^2 - \frac{x^3}{2} + \frac{x^4}{16}\right)dx = \left[\frac{x^3}{3} - \frac{x^4}{8} + \frac{x^5}{80}\right]_0^4\\ &= \frac{63}{3} - 32 + \frac{64}{5} = \frac{32}{15}.\end{aligned}$$

**9.** $\int_C (y - x)\,dx + (2x - y)\,dy$

$C$: boundary of the region lying inside the rectangle with vertices

$(5, 3), (-5, 3), (-5, -3)$, and $(5, -3)$

and outside the square with vertices

$(1, 1), (-1, 1), (-1, -1)$, and $(1, -1)$.

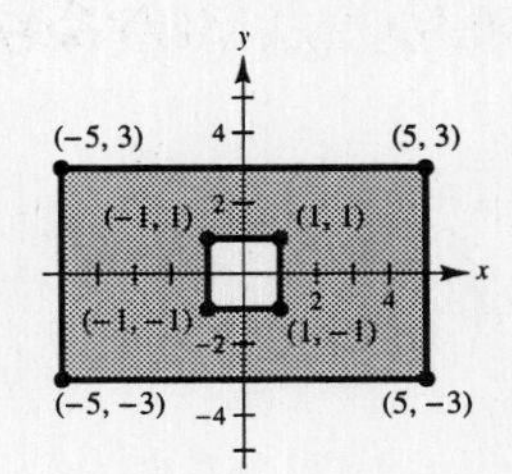


Using $M(x, y) = y - x$, $N(x, y) = 2x - y$, and the figure, we have

$$\begin{aligned}\int_C (y - x)\,dx + (2x - y)\,dy &= \int_R\!\!\int\left(\frac{\partial N}{\partial x} - \frac{\partial M}{\partial y}\right)dA\\ &= \int_R\!\!\int (2 - 1)\,dA\\ &= \text{area of region}\\ &= (\text{area of rectangle}) - (\text{area of square})\\ &= 6(10) - 2(2) = 56.\end{aligned}$$

**15.** $\displaystyle\int_C 2\arctan\frac{y}{x}\,dx + \ln(x^2+y^2)\,dy$

$C$: $x = 4 + 2\cos\theta$, $y = 4 + \sin\theta$

By Green's Theorem, we have

$$\int_C M\,dx + N\,dy = \int_C 2\arctan\frac{y}{x}\,dx + \ln(x^2+y^2)\,dy$$
$$= \int_R\int\left(\frac{\partial N}{\partial x} - \frac{\partial M}{\partial y}\right)dA$$
$$= \int_R\int\left[\frac{2x}{x^2+y^2} - \frac{2(1/x)}{1+(y/x)^2}\right]dA$$
$$= \int_R\int\left[\frac{2x}{x^2+y^2} - \frac{2}{x^2+y^2}\right]dA = 0.$$

**21.** $\mathbf{F}(x, y) = xy\mathbf{i} + (x+y)\mathbf{j}$

$C$: $x^2 + y^2 = 4$

$$\text{Work} = \int_C xy\,dx + (x+y)\,dy$$
$$= \int_R\int\left(\frac{\partial N}{\partial x} - \frac{\partial M}{\partial y}\right)dA = \int_R\int(1-x)\,dA$$
$$= \int_0^{2\pi}\int_0^2(1 - r\cos\theta)r\,dr\,d\theta$$
$$= \int_0^{2\pi}\left[\frac{1}{2}r^2 - \frac{1}{3}r^3\cos\theta\right]_0^2 d\theta$$
$$= \int_0^{2\pi}\left(2 - \frac{8}{3}\cos\theta\right)d\theta = \left[2\theta - \frac{8}{3}\sin\theta\right]_0^{2\pi} = 4\pi$$

**27.** $R$: $y = 2x + 1$ and $y = 4 - x^2$

The region $R$ (see figure) is enclosed by the path $C$ given by

$C_1$: $x = t$, $y = 2t + 1$, $\quad dx = dt$, $dy = 2\,dt$, $\quad -3 \le t \le 1$

$C_2$: $x = 1 - t$, $y = 4 - (1-t)^2 = 3 + 2t - t^2$, $\quad dx = -dt$, $dy = (2 - 2t)\,dt$, $\quad 0 \le t \le 4$.

Therefore, the area of $R$ is

$$A = \frac{1}{2}\int_C x\,dy - y\,dx$$
$$= \frac{1}{2}\int_{C_1} t(2\,dt) - (2t-1)\,dt + \frac{1}{2}\int_{C_2}(1-t)(2-2t)\,dt - (3+2t-t^2)(-dt)$$
$$= \frac{1}{2}\int_{-3}^{1}(-1)\,dt + \frac{1}{2}\int_0^4(5 - 2t + t^2)\,dt$$
$$= \left[-\frac{t}{2}\right]_{-3}^{1} + \frac{1}{2}\left[5t - t^2 + \frac{t^3}{3}\right]_0^4$$
$$= -2 + \frac{1}{2}\left(20 - 16 + \frac{64}{3}\right) = \frac{32}{3}.$$

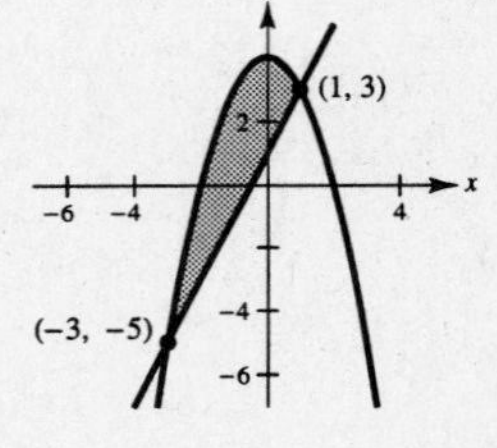


**33.** $R$: $y = x^3$, $y = x$, $0 \le x \le 1$

The region $R$ (see figure) enclosed by the path $C$ given by

$C_1$: $x = t$, $y = t^3$, $\quad dx = dt$, $dy = 3t^2\,dy$, $\quad 0 \le t \le 1$

$C_2$: $x = 1 - t$, $y = 1 - t$, $\quad dx = -dt$, $dy = -dt$, $\quad 0 \le t \le 1$.

Therefore, the area of $R$ is

$$A = \frac{1}{2}\int_C x\,dy - y\,dx$$
$$= \frac{1}{2}\int_{C_1} t(3t^2)\,dt - t^3\,dt + \frac{1}{2}\int_{C_2}(1-t)(-dt) - (1-t)(-dt)$$
$$= \frac{1}{2}\int_0^1 2t^3\,dt = \left[\frac{1}{4}t^4\right]_0^1 = \frac{1}{4}.$$

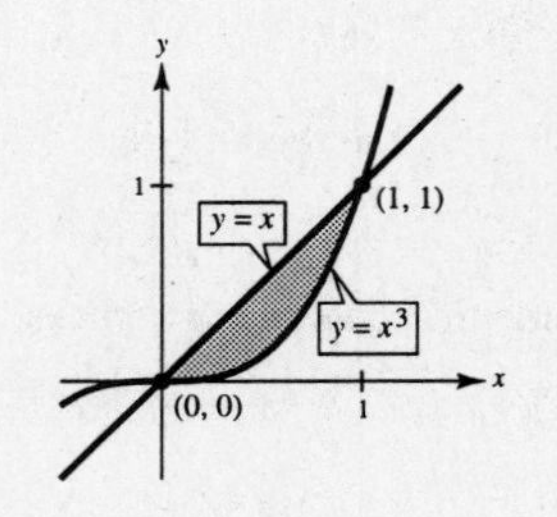


—CONTINUED—

**33. —CONTINUED—**

Thus,

$$\bar{x} = \frac{1}{2A}\int_C x^2\,dy = 2\left[\int_{C_1} t^2(3t^2)\,dt + \int_{C_2}(1-t)^2(-dt)\right] = 2\int_0^1 [3t^4 - (1-t)^2]\,dt$$

$$= 2\left[\frac{3}{5}t^5 + \frac{1}{3}(1-t)^3\right]_0^1 = 2\left[\frac{3}{5} - \frac{1}{3}\right] = \frac{8}{15}$$

and

$$\bar{y} = -\frac{1}{2A}\int_C y^2\,dx = -2\left[\int_{C_1} t^6\,dt + \int_{C_2}(1-t)^2(-dt)\right] = -2\int_0^1 [t^6 - (1-t)^2]\,dt$$

$$= -2\left[\frac{1}{7}t^7 + \frac{1}{3}(1-t)^3\right]_0^1 = -2\left[\frac{1}{7} - \frac{1}{3}\right] = \frac{8}{21}.$$

**37.** $R$: $r = 1 + 2\cos\theta$ (inner loop)

The inner loop of $r = 1 + 2\cos\theta$ starts at $\theta = 2\pi/3$ and ends at $\theta = 4\pi/3$ (see figure). Hence the area enclosed by this inner loop is

$$A = \frac{1}{2}\int_{2\pi/3}^{4\pi/3} (1 + 2\cos\theta)^2\,d\theta$$

$$= \frac{1}{2}\int_{2\pi/3}^{4\pi/3}\left[1 + 4\cos\theta + 4\left(\frac{1+\cos 2\theta}{2}\right)\right]d\theta$$

$$= \frac{1}{2}\int_{2\pi/3}^{4\pi/3} (3 + 4\cos\theta + 2\cos 2\theta)\,d\theta$$

$$= \frac{1}{2}\left[3\theta + 4\sin\theta + \sin 2\theta\right]_{2\pi/3}^{4\pi/3} = \pi - \frac{3\sqrt{3}}{2}.$$

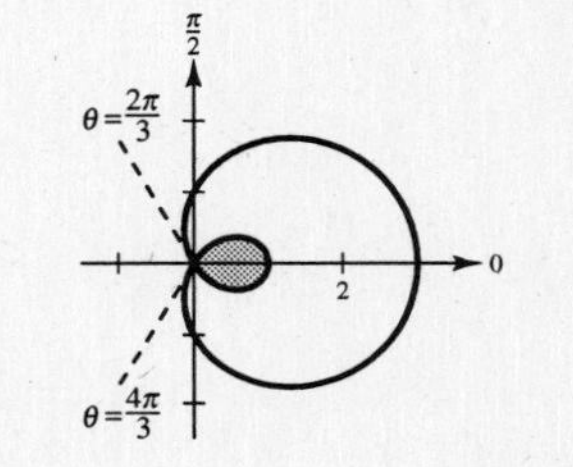


## Section 14.5 Parametric Surfaces

**7.** $\mathbf{r}(u, v) = 2\cos u\mathbf{i} + v\mathbf{j} + 2\sin u\mathbf{k}$

To identify the surface, we can use the trigonometric identity $\sin^2\theta + \cos^2\theta = 1$ to eliminate the parameter $u$ and obtain

$$x^2 + z^2 = (2\cos u)^2 + (2\sin u)^2 = 4.$$

Since $y = v$ can be any real number, the surface is a circular cylinder of radius 2 and rulings parallel to the $y$-axis.

**11.** $\mathbf{s}(u, v) = u\cos v\mathbf{i} + u\sin v\mathbf{j} + u^2\mathbf{k}$

$0 \le u \le 3, 0 \le v \le 2\pi$

For a specific value of $u$ in the interval $0 \le u \le 3$, we have a circular trace of radius $u$ located $u^2$ units above the $xy$-plane. Therefore, $\mathbf{s}$ differs from $\mathbf{R}$ in that the height of the paraboloid is increased from 4 to 9.

**21.** (a) $\mathbf{r}(u, v) = (4 + \cos v)\cos u\mathbf{i} + (4 + \cos v)\sin u\mathbf{j} + \sin v\mathbf{k}$

$0 \le u \le 2\pi, 0 \le v \le 2\pi$

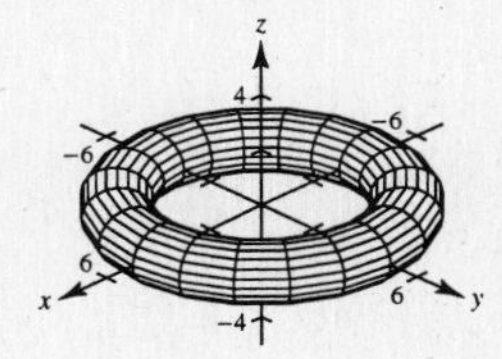


**—CONTINUED—**

**21. —CONTINUED—**

(b) $\mathbf{r}(u, v) = (4 + 2\cos v)\cos u\mathbf{i} + (4 + 2\cos v)\sin u\mathbf{j} + 2\sin v\mathbf{k}$

$0 \le u \le 2\pi, 0 \le v \le 2\pi$

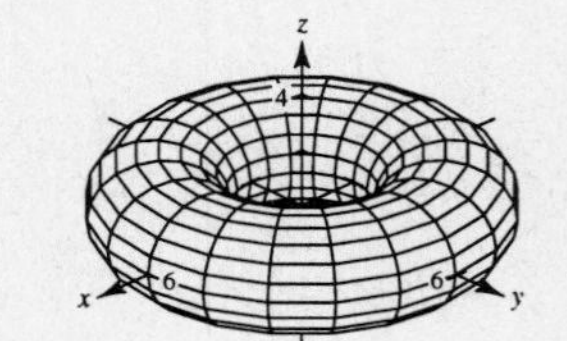


(c) $\mathbf{r}(u, v) = (8 + \cos v)\cos u\mathbf{i} + (8 + \cos v)\sin u\mathbf{j} + \sin v\mathbf{k}$

$0 \le u \le 2\pi, 0 \le v \le 2\pi$

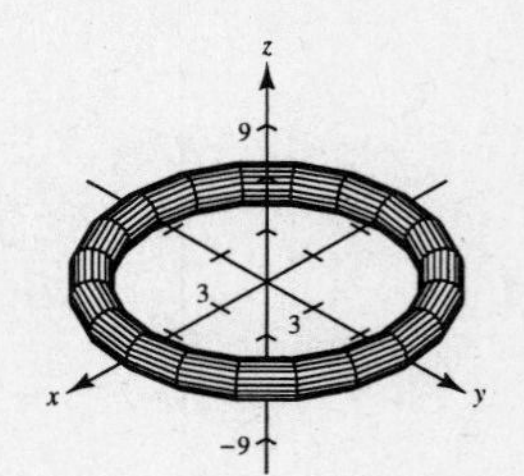


(d) $\mathbf{r}(u, v) = (8 + 3\cos v)\cos u\mathbf{i} + (8 + 3\cos v)\sin u\mathbf{j} + 3\sin v\mathbf{k}$

$0 \le u \le 2\pi, 0 \le v \le 2\pi$

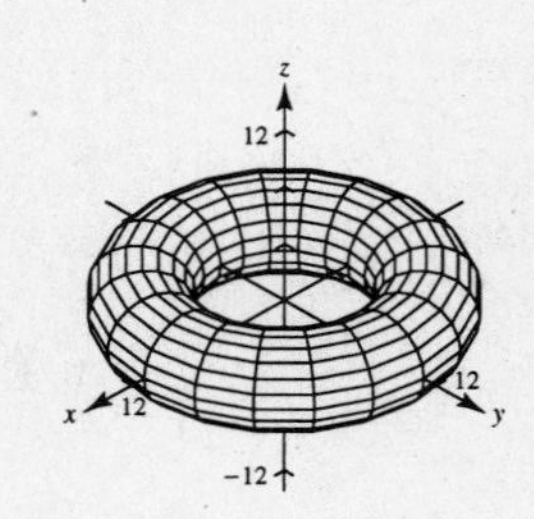


**27.** The graph is a cylinder whose generating curve, $z = x^2$, is a parabola in the $xz$-plane. The rulings of the cylinder are parallel to the $y$-axis (see figure). Therefore, if $x = u$, then $z = u^2$. Since there is no restriction on $y$, we let $y = v$. Thus, the vector-valued function is

$$\mathbf{r}(u, v) = u\mathbf{i} + v\mathbf{j} + u^2\mathbf{k}.$$

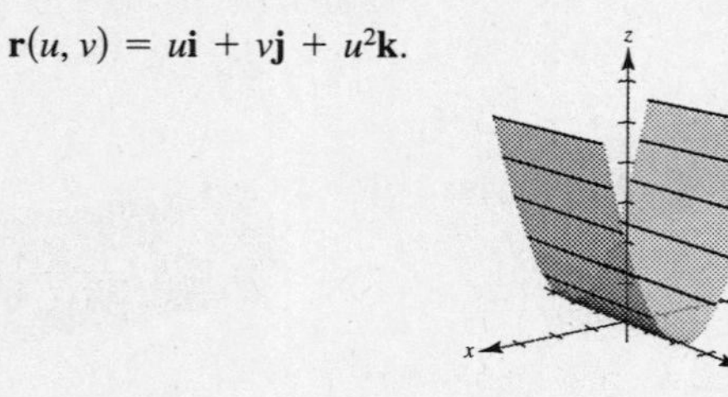


**33.** Surface: $x = f(z) = \sin z, 0 \le z \le \pi$ revolved about the $z$-axis. If we use the parameters $u$ and $v$ and let $x = f(u)\cos v$ and $y = f(u)\sin v$, then

$$x^2 + y^2 = [f(u)\cos v]^2 + [f(u)\sin v]^2 = [f(u)]^2.$$

Therefore, a parametric representation of the surface of revolution is

$$x = \sin u \cos v, \; y = \sin u \sin v, \text{ and } z = u$$

where $0 \le u \le \pi$ and $0 \le v \le 2\pi$.

**37.** $\mathbf{r}(u, v) = 2u\cos v\mathbf{i} + 3u\sin v\mathbf{j} + u^2\mathbf{k}$

The point in the $uv$-plane that is mapped to the point $(x, y, z) = (0, 6, 4)$ is $(u, v) = (2, \pi/2)$. The partial derivatives of $\mathbf{r}$ are $\mathbf{r}_u = 2\cos v\mathbf{i} + 3\sin v\mathbf{j} + 2u\mathbf{k}$ and $\mathbf{r}_v = -2u\sin v\mathbf{i} + 3u\cos v\mathbf{j}$. Therefore,

$$\mathbf{r}_u\left(2, \frac{\pi}{2}\right) = 3\mathbf{j} + 4\mathbf{k} \quad \text{and} \quad \mathbf{r}_v\left(2, \frac{\pi}{2}\right) = -4\mathbf{i}.$$

The normal vector at the point $(0, 6, 4)$ on the surface is

$$\mathbf{r}_u \times \mathbf{r}_v = \begin{vmatrix} \mathbf{i} & \mathbf{j} & \mathbf{k} \\ 0 & 3 & 4 \\ -4 & 0 & 0 \end{vmatrix} = -16\mathbf{j} + 12\mathbf{k}.$$

Thus, an equation of the tangent plane at $(0, 6, 4)$ is

$$0(x - 0) - 16(y - 6) + 12(z - 4) = 0$$

$$4y - 3z = 12.$$

**43.** $\mathbf{r}(u, v) = au \cos v\mathbf{i} + au \sin v\mathbf{j} + u\mathbf{k}$

$0 \le u \le b, 0 \le v \le 2\pi$

Begin by calculating $\mathbf{r}_u$ and $\mathbf{r}_v$.

$$\mathbf{r}_u = a \cos v\mathbf{i} + a \sin v\mathbf{j} + \mathbf{k}$$

$$\mathbf{r}_v = -au \sin v\mathbf{i} + au \cos v\mathbf{j}$$

The cross product of these two vectors is

$$\mathbf{r}_u \times \mathbf{r}_v = \begin{vmatrix} \mathbf{i} & \mathbf{j} & \mathbf{k} \\ a\cos v & a \sin v & 1 \\ -au \sin v & au \cos v & 0 \end{vmatrix} = -au \cos v\mathbf{i} - au \sin v\mathbf{j} + a^2u\mathbf{k}$$

which implies that

$$\begin{aligned} \|\mathbf{r}_u \times \mathbf{r}_v\| &= \sqrt{(-au \cos v)^2 + (-au \sin v)^2 + (a^2u)^2} \\ &= \sqrt{a^2u^2 + a^4u^2} \\ &= au\sqrt{1 + a^2} \quad (0 < a, 0 \le u). \end{aligned}$$

Finally, the surface area of the specified portion of the cone is

$$\begin{aligned} A = \int_R\int \|\mathbf{r}_u \times \mathbf{r}_v\| &= \int_0^{2\pi}\int_0^b au\sqrt{1 + a^2}\, du\, dv \\ &= a\sqrt{1 + a^2}\int_0^{2\pi} \frac{b^2}{2}\, dv \\ &= \pi ab^2\sqrt{1 + a^2}. \end{aligned}$$

## Section 14.6 Surface Integrals

**7.** $\displaystyle\int_S\int xy\, dS$

$S$: $z = 9 - x^2$ for $0 \le x \le 2$, and $0 \le y \le x$

Begin by writing the equation for the surface $S$ as $z = g(x, y) = 9 - x^2$ so that $g_x(x, y) = -2x$ and $g_y(x, y) = 0$, to obtain

$$\sqrt{1 + [g_x(x, y)]^2 + [g_y(x, y)]^2} = \sqrt{1 + 4x^2}.$$

Using the figure, we have

$$\begin{aligned} \int_S\int xy\, dS &= \int_R\int f(x, y, g(x, y))\sqrt{1 + [g_x(x, y)]^2 + [g_y(x, y)]^2}\, dA \\ &= \int_0^2\int_0^x xy\sqrt{1 + 4x^2}\, dy\, dx \\ &= \frac{1}{2}\int_0^2 x^3\sqrt{1 + 4x^2}\, dx. \end{aligned}$$

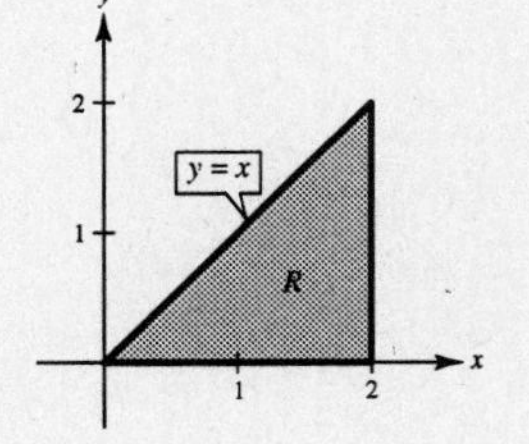


Using a symbolic integration utility to evaluate the last integral yields

$$\int_S\int xy\, dS = \frac{391\sqrt{17} + 1}{240}.$$

If we do not have access to a symbolic integration utility, use trigonometric substitution letting $2x = \tan\theta$ and $2\, dx = \sec^2\theta\, d\theta$ to obtain

$$\begin{aligned} \int_S\int xy\, dS &= \frac{1}{32}\int_0^{\arctan 4} \tan^3\theta \sec^3\theta\, d\theta \\ &= \frac{1}{32}\left[\frac{1}{5}\sec^5\theta - \frac{1}{3}\sec^3\theta\right]_0^{\arctan 4} = \frac{391\sqrt{17} + 1}{240}. \end{aligned}$$

**17.** $\displaystyle\iint_S \sqrt{x^2+y^2+z^2}\,dS$

$S$: $z = \sqrt{x^2+y^2}$ for $x^2+y^2 \le 4$

Begin by writing the equation of the surface $S$ as $z = g(x, y) = \sqrt{x^2+y^2}$, so that

$$g_x(x, y) = \frac{x}{\sqrt{x^2+y^2}} \quad \text{and} \quad g_y(x, y) = \frac{y}{\sqrt{x^2+y^2}}$$

to obtain

$$\sqrt{1+[g_x(x,y)]^2+[g_y(x,y)]^2} = \sqrt{1+\frac{x^2}{x^2+y^2}+\frac{y^2}{x^2+y^2}} = \sqrt{2}.$$

Using the figure, we have

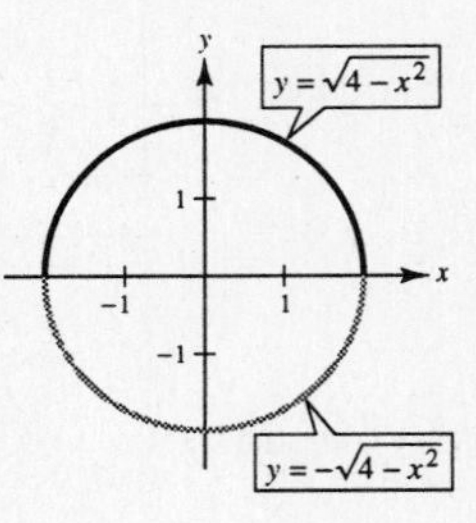


$$\begin{aligned}
\iint_S \sqrt{x^2+y^2+z^2}\,dS &= \iint_R f(x, y, g(x, y))\sqrt{1+[g_x(x,y)]^2+[g_y(x,y)]^2}\,dA \\
&= \int_{-2}^{2}\int_{-\sqrt{4-x^2}}^{\sqrt{4-x^2}} \sqrt{x^2+y^2+\left(\sqrt{x^2+y^2}\right)^2}\sqrt{2}\,dy\,dx \\
&= 2\int_{-2}^{2}\int_{-\sqrt{4-x^2}}^{\sqrt{4-x^2}} \sqrt{x^2+y^2}\,dy\,dx \\
&= 2\int_0^{2\pi}\int_0^2 r(r\,dr\,d\theta) \qquad \text{(polar coordinates)} \\
&= 2\int_0^{2\pi}\left[\frac{r^3}{3}\right]_0^2 d\theta = \frac{16}{3}\int_0^{2\pi} d\theta = \frac{32\pi}{3}.
\end{aligned}$$

**23.** $\mathbf{F}(x, y, z) = x\mathbf{i} + y\mathbf{j} + z\mathbf{k}$

$S$: $z = 9 - x^2 - y^2, 0 \le z$

The vector field $\mathbf{F}$, over the surface $S$, is given by

$$\mathbf{F}(x, y, z) = x\mathbf{i} + y\mathbf{j} + z\mathbf{k} = x\mathbf{i} + y\mathbf{j} + (9 - x^2 - y^2)\mathbf{k}.$$

We write the equation for the surface $S$ as $z = g(x, y) = 9 - x^2 - y^2$ so that

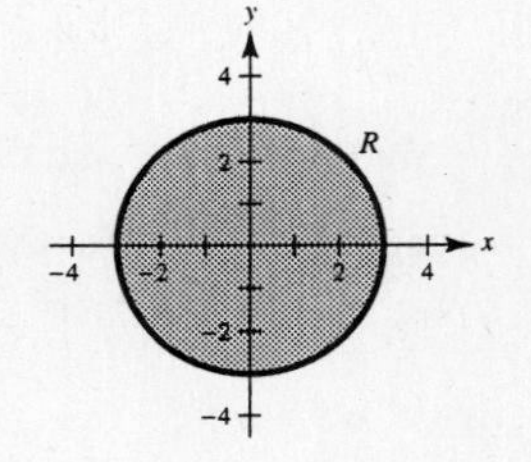


$$g_x(x, y) = -2x \quad \text{and} \quad g_y(x, y) = -2y.$$

$$\begin{aligned}
\iint_S \mathbf{F}\cdot\mathbf{N}\,dS &= \iint_R \mathbf{F}\cdot[-g_x(x,y)\mathbf{i} - g_y(x,y)\mathbf{j} + \mathbf{k}]\,dA \\
&= \iint_R [x\mathbf{i} + y\mathbf{j} + (9 - x^2 - y^2)\mathbf{k}]\cdot(2x\mathbf{i} + 2y\mathbf{j} + \mathbf{k})\,dA \\
&= \iint_R (9 + x^2 + y^2)\,dA \\
&= 4\int_0^{\pi/2}\int_0^3 (9 + r^2)r\,dr\,d\theta \qquad \text{(polar coordinates)} \\
&= \frac{243\pi}{2}
\end{aligned}$$

**33.** $S$: $x^2 + y^2 = a^2 (0 \le z \le h)$

Note that $S$ does not define $z$ as a function of $x$ and $y$. Hence, project onto the $xz$-plane, so that $y = \sqrt{a^2 - x^2} = g(x, z)$ and obtain

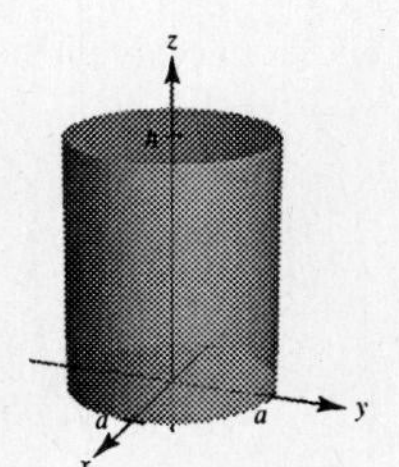


$$\sqrt{1+[g_x(x,z)]^2+[g_z(x,z)]^2} = \sqrt{1+\frac{x^2}{a^2-x^2}} = \frac{a}{\sqrt{a^2-x^2}}.$$

—CONTINUED—

**33. —CONTINUED—**

$$I_z = \iint_S (x^2 + y^2)(1)\,dS$$

$$= \iint_R a^2\sqrt{1 + [g_x(x, z)]^2 + [g_z(x, z)]^2}\,dA$$

$$= 4a^2\int_0^a\int_0^h \frac{a}{\sqrt{a^2 - x^2}}\,dz\,dx$$

$$= 4a^3h\int_0^a \frac{1}{\sqrt{a^2 - x^2}}\,dx$$

$$= 4a^3h\left[\arcsin\frac{x}{a}\right]_0^a = 2\pi a^3h$$

## Section 14.7 Divergence Theorem

**3.** $\mathbf{F}(x, y, z) = (2x - y)\mathbf{i} - (2y - z)\mathbf{j} + z\mathbf{k}$

**Surface Integral**: There are four surfaces to this solid

As shown in the figure, there are four surfaces to the solid bounded by the coordinate planes and $2x + 4y + 2z = 12$.

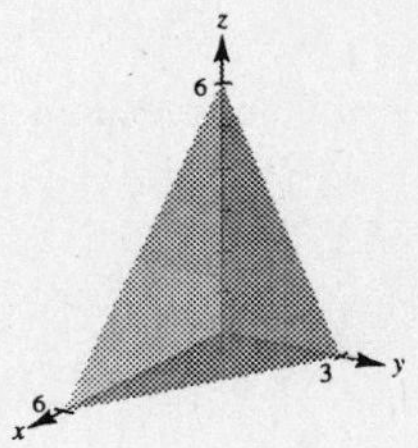

$z = 0, \mathbf{N} = -\mathbf{k}, \mathbf{F}\cdot\mathbf{N} = -z$

$$\iint_{S_1} 0\,dS = 0$$

$y = 0, \mathbf{N} = -\mathbf{j}, \mathbf{F}\cdot\mathbf{N} = 2y - z, dS = dA = dx\,dz$

$$\iint_{S_2} -z\,dS = \int_0^6\int_0^{6-z} -z\,dx\,dz = \int_0^6 (z^2 - 6z)\,dx = -36$$

$x = 0, \mathbf{N} = -\mathbf{i}, \mathbf{F}\cdot\mathbf{N} = y - 2x, dS = dA = dz\,dy$

$$\iint_{S_3} y\,dS = \int_0^3\int_0^{6-2y} y\,dz\,dy = \int_0^3 (6y - 2y^2)\,dy = 9$$

$x + 2y + z = 6, \mathbf{N} = \dfrac{\mathbf{i} + 2\mathbf{j} + \mathbf{k}}{\sqrt{6}}, \mathbf{F}\cdot\mathbf{N} = \dfrac{2x - 5y + 3z}{\sqrt{6}}, dS = \sqrt{6}\,dA$

$$\iint_{S_4} (2x - 5y + 3z)\,dz\,dy = \int_0^3\int_0^{6-2y} (18 - x - 11y)\,dx\,dy = \int_0^3 (90 - 90y + 20y^2)\,dy = 45$$

Therefore, $\displaystyle\iint_S \mathbf{F}\cdot\mathbf{N}\,dS = 0 - 36 + 9 + 45 = 18.$

**Divergence Theorem**: Since div $\mathbf{F} = 1$, we have

$$\iiint_Q dV = \text{(Volume of solid)} = \frac{1}{3}\text{(Area of base)} \times \text{(Height)} = \frac{1}{3}(9)(6) = 18.$$

**9.** $\mathbf{F}(x, y, z) = x\mathbf{i} + y\mathbf{j} + z\mathbf{k}$

$S$: $x^2 + y^2 + z^2 = 4$

Since div $\mathbf{F}(x, y, z) = 1 + 1 + 1 = 3$, it follows that

$$\iiint_Q \text{div}\,\mathbf{F}\,dV = 3\iiint_Q dV$$

$$= 3\text{(volume of sphere of radius 2)}$$

$$= 3\left[\frac{4\pi 2^3}{3}\right] = 32\pi.$$

**19.** $\mathbf{F}(x, y, z) = (4xy + z^2)\mathbf{i} + (2x^2 + 6yz)\mathbf{j} + 2xz\mathbf{k}$

$S$: The closed surface of the solid bounded by the graphs of $x = 4$, $z = 9 - y^2$, and the coordinate planes.

Using the Divergence Theorem, we have

$$\int_S\int \mathbf{curl\ F} \cdot \mathbf{N}\, dS = \iiint_Q \operatorname{div}(\mathbf{curl\ F})\, dV.$$

$$\mathbf{curl\ F}(x, y, z) = \begin{vmatrix} \mathbf{i} & \mathbf{j} & \mathbf{k} \\ \dfrac{\partial}{\partial x} & \dfrac{\partial}{\partial y} & \dfrac{\partial}{\partial z} \\ 4xy + z^2 & 2x^2 + 6yz & 2xz \end{vmatrix} = -6y\mathbf{i} - (2z - 2z)\mathbf{j} + (4x - 4x)\mathbf{k} = -6y\mathbf{i}.$$

Therefore, div $(\mathbf{curl\ F}(x, y, z)) = 0$ and

$$\int_S\int \mathbf{curl\ F} \cdot \mathbf{N}\, dS = \iiint_Q \operatorname{div}(\mathbf{curl\ F})\, dV = 0.$$

## Section 14.8 Stoke's Theorem

**9.** $\mathbf{F}(x, y, z) = xyz\mathbf{i} + y\mathbf{j} + z\mathbf{k}$

$S$: $3x + 4y + 2z = 12$ first octant

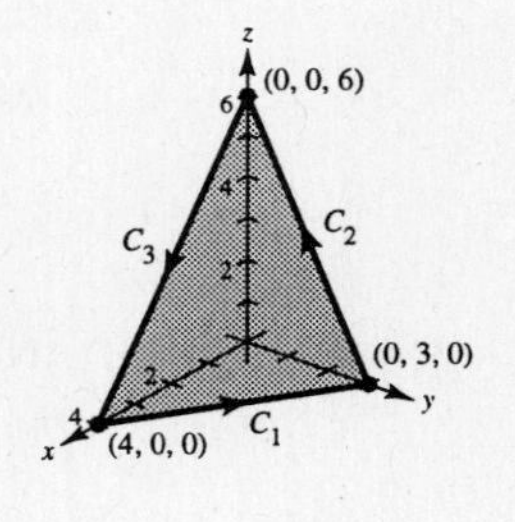


As a *line integral*, integrate along the three paths shown in the figure and obtain

$$\int_{C_1} \mathbf{F} \cdot \mathbf{T}\, ds = \int_C xyz\, dx + y\, dy + z\, dz$$

$$= \int_{C_1} 0\, dx + y\, dy + 0\, dz + \int_{C_2} 0\, dx + y\, dy + z\, dz + \int_{C_3} 0\, dx + 0\, dy + z\, dz$$

$$= \int_0^3 y\, dy + \int_3^0 y\, dy + \int_0^6 z\, dz + \int_6^0 z\, dz = 0.$$

As a *double integral*, begin by finding **curl F**.

$$\mathbf{curl\ F}(x, y, z) = \begin{vmatrix} \mathbf{i} & \mathbf{j} & \mathbf{k} \\ \dfrac{\partial}{\partial x} & \dfrac{\partial}{\partial y} & \dfrac{\partial}{\partial z} \\ xyz & y & z \end{vmatrix} = xy\mathbf{j} - xz\mathbf{k}.$$

The upward normal is $\mathbf{N} = 3\mathbf{i} + 4\mathbf{j} + 2\mathbf{k}$.

$$\int_S\int (\mathbf{curl\ F}) \cdot \mathbf{N}\, dS = \int_R\int (xy\mathbf{j} - xz\mathbf{k}) \cdot (3\mathbf{i} + 4\mathbf{j} + 2\mathbf{k})\, dA$$

$$= \int_R\int (4xy - 2xz)\, dA$$

$$= \int_0^4\int_0^{3(4-x)/4} \left[4xy - 2x\left(6 - 2y - \frac{3x}{2}\right)\right] dy\, dx$$

$$= \int_0^4\int_0^{3(4-x)/4} (8xy + 3x^2 - 12x)\, dy\, dx$$

$$= \int_0^4 \left(36x - 18x^2 + \frac{9x^3}{4} + 9x^2 - \frac{9x^3}{4} - 36x + 9x^2\right) dx$$

$$= \int_0^4 (0)\, dx = 0.$$

**11.** $\mathbf{F}(x, y, z) = 2y\mathbf{i} + 3z\mathbf{j} - x\mathbf{k}$

$C$: triangle with vertices (0, 0, 0), (0, 2, 0), and (1, 1, 1)

$$\textbf{curl F} = \begin{vmatrix} \mathbf{i} & \mathbf{j} & \mathbf{k} \\ \dfrac{\partial}{\partial x} & \dfrac{\partial}{\partial y} & \dfrac{\partial}{\partial z} \\ 2y & 3z & -x \end{vmatrix} = (0 - 3)\mathbf{i} - (-1 - 0)\mathbf{j} + (0 - 2)\mathbf{k} = -3\mathbf{i} + \mathbf{j} - 2\mathbf{k}$$

Using the coordinates of the vertices of the triangle we obtain the vectors **u** and **v** forming two of its edges. They are

$$\mathbf{u} = \mathbf{i} + \mathbf{j} + \mathbf{k} \quad \text{and} \quad \mathbf{v} = 0\mathbf{i} + 2\mathbf{j} + 0\mathbf{k} = 2\mathbf{j}.$$

Therefore, a vector normal to the surface is given by

$$\mathbf{u} \times \mathbf{v} = \begin{vmatrix} \mathbf{i} & \mathbf{j} & \mathbf{k} \\ 1 & 1 & 1 \\ 0 & 2 & 0 \end{vmatrix} = -2\mathbf{i} + 2\mathbf{k},$$

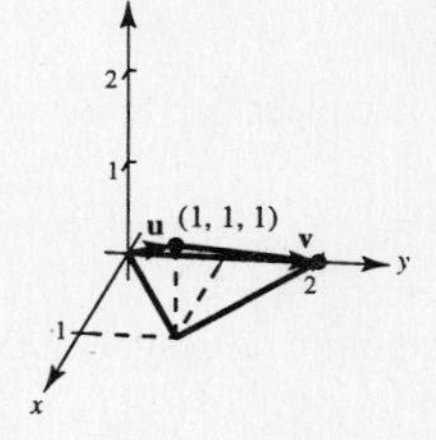


and a unit vector normal to the surface is

$$\mathbf{N} = \frac{\mathbf{u} \times \mathbf{v}}{\|\mathbf{u} \times \mathbf{v}\|} = \frac{-2\mathbf{i} + 2\mathbf{k}}{2\sqrt{2}} = \frac{-\mathbf{i} + \mathbf{k}}{\sqrt{2}}.$$

Thus, the surface (plane) is given by $f(x, y, z) = -x + z$, and we have $dS = \sqrt{1 + 1}\, dA = \sqrt{2}\, dA$. We conclude that

$$\int_C \mathbf{F} \cdot d\mathbf{r} = \iint_S (\textbf{curl F}) \cdot \mathbf{N}\, dS$$

$$= \iint_R \frac{(3 - 2)}{\sqrt{2}} \sqrt{2}\, dA$$

$$= \iint_R dA = \text{area of triangle} = \left(\frac{1}{2}\right)(2)(1) = 1.$$

**17.** $\mathbf{F}(x, y, z) = -\ln\sqrt{x^2 + y^2}\,\mathbf{i} + \arctan\dfrac{x}{y}\mathbf{j} + \mathbf{k}$

$S$: first octant portion of the plane $z = 9 - 2x - 3y$ over one petal of the rose curve $r = 2 \sin 2\theta$

$$\textbf{curl F} = \begin{vmatrix} \mathbf{i} & \mathbf{j} & \mathbf{k} \\ \dfrac{\partial}{\partial x} & \dfrac{\partial}{\partial y} & \dfrac{\partial}{\partial z} \\ -\ln\sqrt{x^2 + y^2} & \arctan\dfrac{x}{y} & 1 \end{vmatrix} = \left[\frac{1/y}{1 + (x^2/y^2)} + \frac{y}{x^2 + y^2}\right]\mathbf{k} = \left[\frac{2y}{x^2 + y^2}\right]\mathbf{k}.$$

Since $S$ is the first octant portion of the plane $z = 9 - 2x - 3y$ over one petal of $r = 2 \sin 2\theta$, we have

$$\mathbf{N} = \frac{2\mathbf{i} + 3\mathbf{j} + \mathbf{k}}{\sqrt{14}} \quad \text{and} \quad dS = \sqrt{1 + (-2)^2 + (-3)^2}\, dA = \sqrt{14}\, dA.$$

Therefore,

$$\iint_S \textbf{curl F} \cdot \mathbf{N}\, dS = \iint_R \frac{2y}{x^2 + y^2} \frac{1}{\sqrt{14}} \sqrt{14}\, dA$$

$$= \iint_R \frac{2y}{x^2 + y^2}\, dA$$

$$= \int_0^{\pi/2} \int_0^{2 \sin 2\theta} \frac{2r \sin \theta}{r^2} r\, dr\, d\theta \qquad \text{(polar coordinates)}$$

$$= \int_0^{\pi/2} \int_0^{4 \sin \theta \cos \theta} 2 \sin \theta\, dr\, d\theta$$

$$= \int_0^{\pi/2} 8 \sin^2 \theta \cos \theta\, d\theta = \left[\frac{8 \sin^3 \theta}{3}\right]_0^{\pi/2} = \frac{8}{3}.$$

## Review Exercises for Chapter 14

**11.** $\mathbf{F}(x, y, z) = \dfrac{yz\mathbf{i} - xz\mathbf{j} - xy\mathbf{k}}{y^2z^2}$

$$\textbf{curl } \mathbf{F}(x, y, z) = \begin{vmatrix} \mathbf{i} & \mathbf{j} & \mathbf{k} \\ \dfrac{\partial}{\partial x} & \dfrac{\partial}{\partial y} & \dfrac{\partial}{\partial z} \\ \dfrac{1}{yz} & \dfrac{-x}{y^2z} & \dfrac{-x}{yz^2} \end{vmatrix} = \left(\frac{x}{y^2z^2} - \frac{x}{y^2z^2}\right)\mathbf{i} - \left(\frac{-1}{yz^2} - \frac{-1}{yz^2}\right)\mathbf{j} + \left(\frac{-1}{y^2z} - \frac{-1}{y^2z}\right)\mathbf{k} = 0$$

Therefore, $\mathbf{F}$ is conservative. Now, if $f$ is a function such that $\mathbf{F}(x, y, z) = \nabla f(x, y, z)$, then

$$f_x(x, y, z) = \frac{1}{yz}, \; f_y(x, y, z) = -\frac{x}{y^2z}, \text{ and } f_z(x, y, z) = -\frac{x}{yz^2}$$

and by integrating with respect to $x$, $y$, and $z$ separately, we obtain

$$f(x, y, z) = \int \frac{1}{yz}\,dx = \frac{x}{yz} + g(y, z) + K$$

$$f(x, y, z) = \int -\frac{x}{y^2z}\,dy = \frac{x}{yz} + h(x, z) + K$$

$$f(x, y, z) = \int -\frac{x}{yz^2}\,dz = \frac{x}{yz} + k(x, y) + K.$$

By comparing these three versions of $f(x, y, z)$, we can conclude that $g(y, z) = h(x, z) = k(x, y) = 0$, and

$$f(x, y, z) = \frac{x}{yz} + K.$$

**15.** $\mathbf{F}(x, y, z) = (\cos y + y\cos x)\mathbf{i} + (\sin x - x\sin y)\mathbf{j} + xyz\mathbf{k}$.

(a) $$\text{div } \mathbf{F}(x, y, z) = \frac{\partial}{\partial x}[\cos y + y\cos x] + \frac{\partial}{\partial y}[\sin x - x\sin y] + \frac{\partial}{\partial z}[xyz]$$
$$= -y\sin x - x\cos y + xy$$

(b) $$\textbf{curl } \mathbf{F}(x, y, z) = \begin{vmatrix} \mathbf{i} & \mathbf{j} & \mathbf{k} \\ \dfrac{\partial}{\partial x} & \dfrac{\partial}{\partial y} & \dfrac{\partial}{\partial z} \\ \cos y + y\cos x & \sin x - x\sin y & xyz \end{vmatrix}$$
$$= (xz - 0)\mathbf{i} - (yz - 0)\mathbf{j} + (\cos x - \sin y + \sin y - \cos x)\mathbf{k}$$
$$= xz\mathbf{i} - yz\mathbf{j}$$

**25.** $\displaystyle\int_C (2x - y)\,dx + (x + 3y)\,dy$

(a) $C$ is the line segment from $(0, 0)$ to $(2, -3)$.

$$C: x = t,\; dx = dt,\; y = -\frac{3t}{2},\; dy = \left(-\frac{3}{2}\right)dt,\; 0 \le t \le 2$$

Therefore,

$$\int_C (2x - y)\,dx + (x + 3y)\,dy = \int_0^2 \left[\frac{7t}{2}\,dt + \left(-\frac{7t}{2}\right)\left(-\frac{3}{2}\,dt\right)\right]$$
$$= \int_0^2 \frac{35}{4}t\,dt = \left[\frac{35}{8}t^2\right]_0^2 = \frac{35}{2}.$$

—CONTINUED—

**25. —CONTINUED—**

(b) $C$ is one counterclockwise revolution on the circle $x = 3\cos t$ and $y = 3\sin t$.

$C$: $x = 3\cos t$, $dx = -3\sin t\,dt$, $y = 3\sin t$, $dy = 3\cos t\,dt$, $0 \le t \le 2\pi$

Therefore,

$$\int_C (2x - y)\,dx + (x + 3y)\,dy = \int_0^{2\pi} [(6\cos t - 3\sin t)(-3\sin t) + (3\cos t + 9\sin t)(3\cos t)]\,dt$$

$$= \int_0^{2\pi} (9\sin t\cos t + 9)\,dt = \left[\frac{9\sin^2 t}{2} + 9t\right]_0^{2\pi} = 18\pi.$$

**29.** $f(x, y) = 5 + \sin(x + y)$

$C$: $y = 3x$ from $(0, 0)$ to $(2, 6)$

A vector-valued function for the path $C$ is

$$\mathbf{r}(t) = t\mathbf{i} + 3t\mathbf{j},\ 0 \le t \le 2,$$

and

$$f(x(t), y(t)) = 5 + \sin(x + y) = 5 + \sin 4t.$$

$$ds = \sqrt{[x'(t)]^2 + [y'(t)]^2}\,dt = \sqrt{(1)^2 + (3)^2} = \sqrt{10}$$

Therefore,

$$\text{area} = \int_C f(x, y)\,ds$$

$$= \int_0^2 (5 + \sin 4t)\sqrt{10}\,dt$$

$$= \sqrt{10}\left[5t - \frac{1}{4}\cos 4t\right]_0^2$$

$$= \frac{\sqrt{10}}{4}(41 - \cos 8) \approx 32.528.$$

**35.** $\mathbf{F}(x, y, z) = (y - z)\mathbf{i} + (z - x)\mathbf{j} + (x - y)\mathbf{k}$

$C$: curve of the intersection of $z = x^2 + y^2$ and $x + y = 0$ from $(-2, 2, 8)$ to $(2, -2, 8)$

On the curve of intersection $z = x^2 + (-x)^2 = 2x^2$. Hence, $C$ is given by $x = t$, $y = -t$, $z = 2t^2$, $-2 \le t \le 2$, and we have

$$\mathbf{r}(t) = t\mathbf{i} - t\mathbf{j} + 2t^2\mathbf{k} \quad \text{and} \quad \mathbf{r}'(t) = \mathbf{i} - \mathbf{j} + 4t\mathbf{k}.$$

Thus,

$$\mathbf{F}(x, y, z) = (y - z)\mathbf{i} + (z - x)\mathbf{j} + (x - y)\mathbf{k}$$

$$= (-t - 2t^2)\mathbf{i} + (2t^2 - t)\mathbf{j} + 2t\mathbf{k}$$

and

$$\int_C \mathbf{F} \cdot d\mathbf{r} = \int_a^b \mathbf{F}(x(t), y(t), z(t)) \cdot \mathbf{r}'(t)\,dt$$

$$= \int_{-2}^2 (-2t^2 - t - 2t^2 + t + 8t^2)\,dt$$

$$= \int_{-2}^2 4t^2\,dt = \frac{64}{3}.$$

**41.** $\displaystyle\int_C 2xyz\,dx + x^2z\,dy + x^2y\,dz$

Since,

$$\frac{\partial}{\partial y}[x^2y] = x^2 = \frac{\partial}{\partial z}[x^2z]$$

$$\frac{\partial}{\partial x}[x^2y] = 2xy = \frac{\partial}{\partial z}[2xyz]$$

$$\frac{\partial}{\partial x}[x^2z] = 2xz = \frac{\partial}{\partial y}[xyz],$$

the vector field $\mathbf{F}(x, y, z) = 2xyz\mathbf{i} + x^2z\mathbf{j} + x^2y\mathbf{k}$ is conservative. Therefore,

$$f_x(x, y, z) = 2xyz \Rightarrow f(x, y, z) = \int 2xyz\,dx = x^2yz + g(y, z)$$

$$f_y(x, y, z) = x^2z \Rightarrow f(x, y, z) = \int x^2z\,dy = x^2yz + h(x, z)$$

$$f_z(x, y, z) = x^2y \Rightarrow f(x, y, z) = \int x^2y\,dz = x^2yz + k(x, y).$$

Comparing these three versions of the potential function $f$, we conclude that $f(x, y, z) = x^2yz + C$, and by the Fundamental Theorem we have

$$\int_C 2xyz\,dx + x^2z\,dy + x^2y\,dz = f(1, 4, 3) - f(0, 0, 0) = 12.$$

**49.** $\displaystyle\int_C xy\,dx + x^2\,dy$

$C$: boundary of the region between the graphs of $y = x^2$ and $y = x$

By Green's Theorem and the figure, we have

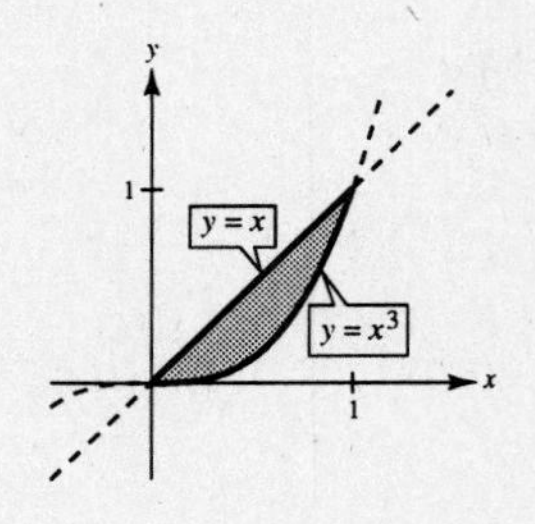


$$\int_C M(x, y)\,dx + N(x, y)\,dy = \int_C xy\,dx + x^2\,dy$$

$$= \iint_R \left[\frac{\partial N}{\partial x} - \frac{\partial M}{\partial y}\right] dA$$

$$= \int_0^1 \int_{x^2}^{x} x\,dy\,dx$$

$$= \int_0^1 (x^2 - x^3)\,dx = \frac{1}{12}.$$

**59.** $\mathbf{F}(x, y, z) = (\cos y + y\cos x)\mathbf{i} + (\sin x - x\sin y)\mathbf{j} + xyz\mathbf{k}$

$S$: portion of $z = y^2$ over the square in the $xy$-plane with vertices $(0, 0)$, $(a, 0)$, $(a, a)$, $(0, a)$

**Line Integral**: Using the line integral we have

$C_1$: $y = 0$, $dy = 0$

$C_2$: $x = 0$, $dx = 0$, $z = y^2$, $dz = 2y\,dy$

$C_3$: $y = a$, $dy = 0$, $z = a^2$, $dz = 0$

$C_4$: $x = a$, $dx = 0$, $z = y^2$, $dz = 2y\,dy$.

$$\int_C \mathbf{F}\cdot d\mathbf{r} = \int_C (\cos y + y\cos x)\,dx + (\sin x - x\sin y)\,dy + xyz\,dz$$

$$= \int_{C_1} dx + \int_{C_2} 0 + \int_{C_3} (\cos a + a\cos x)\,dx + \int_{C_4} (\sin a - a\sin y)\,dy + ay^3(2y\,dy)$$

$$= \int_0^a dx + \int_a^0 (\cos a + a\cos x)\,dx + \int_0^a (\sin a - a\sin y)\,dy + \int_0^a 2ay^4\,dy$$

$$= a + \Big[x\cos a + a\sin x\Big]_a^0 + \Big[y\sin a + a\cos y\Big]_0^a + \left[2a\frac{y^5}{5}\right]_0^a$$

$$= a - a\cos a - a\sin a - a + a\sin a + a\cos a + \frac{2a^6}{5} = \frac{2a^6}{5}$$

**Double Integral**: Consider $f(x, y, z) = y^2 - z$, we have

$$\mathbf{N} = \frac{-\nabla f}{\|\nabla f\|} = \frac{-2y\mathbf{j} + \mathbf{k}}{\sqrt{1 + 4y^2}},\quad dS = \sqrt{1 + 4y^2}\,dA, \text{ and } \mathbf{curl\ F} = xz\mathbf{i} - yz\mathbf{j}.$$

Hence,

$$\iint_S (\mathbf{curl\ F})\cdot\mathbf{N}\,dS = \int_0^a\int_0^a 2y^2z\,dy\,dx = \int_0^a\int_0^a 2y^4\,dy\,dx = \int_0^a \frac{2a^5}{5}\,dx = \frac{2a^6}{5}.$$

# CHAPTER 15
# Differential Equations

# CHAPTER 15
# Differential Equations

## Section 15.1 Exact First-Order Equations

### Solutions to Selected Odd-Numbered Exercises

**3.** $(3y^2 + 10xy^2)\,dx + (6xy - 2 + 10x^2y)\,dy = 0$

Since

$$M(x, y)\,dx + N(x, y)\,dy = (3y^2 + 10xy^2)\,dx + (6xy - 2 + 10x^2y)\,dy = 0,$$

we have

$$\frac{\partial M}{\partial y} = 6y + 20xy = \frac{\partial N}{\partial x}$$

and the equation is exact. We next find a function $f$ such that $f_x(x, y) = M(x, y)$ and $f_y(x, y) = N(x, y)$. The general solution, $f(x, y) = C$, is given by

$$f(x, y) = \int (3y^2 + 10xy^2)\,dx = 3xy^2 + 5x^2y^2 + g(y).$$

Differentiating with respect to $y$ yields

$$f_y(x, y) = 6xy + 10x^2y + g'(y) = \overbrace{6xy - 2 + 10x^2y}^{N(x,\,y)}.$$

Therefore,

$$g'(y) = -2$$

and

$$g(y) = \int -2\,dy = -2y + C_1.$$

Thus, the general solution is

$$f(x, y) = C \quad \text{or} \quad 3xy^2 + 5x^2y^2 - 2y = C.$$

**9.** $\dfrac{1}{(x - y)^2}(y^2\,dx + x^2\,dy) = 0$

Since

$$M(x, y)\,dx + N(x, y)\,dy = \frac{y^2}{(x - y)^2}\,dx + \frac{x^2}{(x - y)^2}\,dy = 0,$$

we have

$$\frac{\partial M}{\partial y} = \frac{(x - y)^2(2y) - y^2(2)(x - y)(-1)}{(x - y)^4} = \frac{2xy}{(x - y)^3}$$

$$\frac{\partial N}{\partial x} = \frac{(x - y)^2(2x) - x^2(2)(x - y)(-1)}{(x - y)^4} = \frac{-2xy}{(x - y)^3}.$$

Since $\dfrac{\partial M}{\partial y} \neq \dfrac{\partial N}{\partial x}$, the differential equation is not exact.

**13.** $\dfrac{y}{x-1}\,dx + [\ln(x-1) + 2y]\,dy = 0$

Since $M(x, y) = \dfrac{y}{x-1}$ and $N(x, y) = [\ln(x-1) + 2y]$, we have

$$\frac{\partial M}{\partial y} = \frac{1}{x-1} = \frac{\partial N}{\partial x}$$

and the differential equation is exact. We next find a function $f$ such that $f_x(x, y) = M(x, y)$ and $f_y(x, y) = N(x, y)$.

$$f(x, y) = \int \frac{y}{x-1}\,dx = y\ln(x-1) + g(y)$$

$$f(x, y) = \int [\ln(x-1) + 2y]\,dy = y\ln(x-1) + y^2 + h(x)$$

Comparing these two versions of $f$, we conclude that $f(x, y) = y\ln(x-1) + y^2$ and the general solution of the differential equation is $y\ln(x-1) + y^2 = C$. Finally, since $y = 4$ when $x = 2$, we have $4\ln 1 + 4^2 = C$ or $C = 16$. Therefore the particular solution is

$$y\ln(x-1) + y^2 = 16.$$

**21.** $(x + y)\,dx + \tan x\,dy = 0$

Since

$$\frac{(\partial M/\partial y) - (\partial N/\partial x)}{N} = \frac{1 - \sec^2 x}{\tan x} = h(x) \text{ and } \int \frac{1 - \sec^2 x}{\tan x}\,dx = -\int \frac{\tan^2 x}{\tan x}\,dx = -\int \tan x\,dx = \ln|\cos x|,$$

it follows that

$$e^{\int h(x)\,dx} = e^{\ln|\cos x|} = \cos x$$

is an integrating factor. Thus,

$$(x + y)\cos x\,dx + \cos x\tan x\,dy = 0$$

$$(x + y)\cos x\,dx + \sin x\,dy = 0$$

is exact. Thus,

$$f(x, y) = \int \sin x\,dy = y\sin x + g(x) \quad \text{and} \quad f_x(x, y) = y\cos x + g'(x) = \overbrace{(x + y)\cos x}^{M(x,\,y)}.$$

Therefore,

$$g'(x) = x\cos x$$

$$g(x) = \int x\cos x\,dx = x\sin x + \cos x + C_1 \qquad \text{Integration by parts}$$

and

$$f(x, y) = y\sin x + x\sin x + \cos x + C_1$$
$$= (x + y)\sin x + \cos x + C_1.$$

Finally, the general solution is

$$f(x, y) = C \quad \text{or} \quad (x + y)\sin x + \cos x = C.$$

**29.** $(-y^5 + x^2y)\,dx + (2xy^4 - 2x^3)\,dy = 0.$

Multiplying the differential equation by the integrating factor $x^{-2}y^{-3}$ yields

$$\left(-\frac{y^2}{x^2} + \frac{1}{y^2}\right) dx + \left(\frac{2y}{x} - \frac{2x}{y^3}\right) dy = 0.$$

Since

$$\frac{\partial M}{\partial y} = -\frac{2y}{x^2} - \frac{2}{y^3} = \frac{\partial N}{\partial x},$$

the differential equation is exact. We next find a function $f$ such that $f_x(x, y) = M(x, y)$ and $f_y(x, y) = N(x, y)$.

$$f(x, y) = \int \left(-\frac{y^2}{x^2} + \frac{1}{y^2}\right) dx = \frac{y^2}{x} + \frac{x}{y^2} + g(y)$$

$$f(x, y) = \int \left(\frac{2y}{x} - \frac{2x}{y^3}\right) dy = \frac{y^2}{x} + \frac{x}{y^2} + h(x)$$

Therefore, $g(y) = h(x) = C_1$ and the general solution to the differential equation is

$$f(x, y) = C \quad \text{or} \quad \frac{y^2}{x} + \frac{x}{y^2} = C.$$

**37.** $\dfrac{dy}{dx} = \dfrac{y - x}{3y - x}$

$(x - y)\,dx + (3y - x)\,dy = 0$

Since $M(x, y) = x - y$ and $N(x, y) = 3y - x$, we have

$$\frac{\partial M}{\partial y} = -1 = \frac{\partial N}{\partial x}$$

and the differential equation is exact. We next find a function $f$ such that $f_x(x, y) = M(x, y)$ and $f_y(x, y) = N(x, y)$.

$$f(x, y) = \int (x - y)\,dx = \frac{1}{2}x^2 - xy + g(y)$$

$$f(x, y) = \int (3y - x)\,dy = \frac{3}{2}y^2 - xy + h(x)$$

Comparing these two versions of $f$, we conclude that $f(x, y) = \frac{1}{2}x^2 - xy + \frac{3}{2}y^2$ and the general solution of the differential equation is $x^2 - 2xy + 3y^2 = C$. Finally, since $y = 1$ when $x = 2$, we have $2^2 - 2(2)(1) + 3(1^2) = C$ or $C = 3$. Therefore the particular solution is

$$x^2 - 2xy + 3y^2 = 3.$$

## Section 15.2 First-Order Linear Differential Equations

**7.** The equation $y' - y = \cos x$ is linear with $P(x) = -1$. Thus,

$$\int P(x)\,dx = -x \quad \text{and} \quad e^{\int P(x)\,dx} = e^{-x}.$$

Multiplying both members of the differential equation in standard form by the integrating factor $e^{-x}$ yields

$$y'e^{-x} - ye^{-x} = e^{-x}\cos x$$

$$\frac{d}{dx}[ye^{-x}] = e^{-x}\cos x$$

$$ye^{-x} = \int e^{-x}\cos x\,dx$$

$$= \frac{1}{2}(e^{-x}\sin x - e^{-x}\cos x) + C$$

$$y = \frac{1}{2}(\sin x - \cos x) + Ce^x.$$

**11.** $(x - 1)y' + y = x^2 - 1.$

In standard form the equation is

$$y' + \left(\frac{1}{x - 1}\right)y = x + 1.$$

Thus,

$$\int P(x)\,dx = \int \frac{1}{x - 1}\,dx = \ln|x - 1|$$

and the integrating factor is

$$e^{\int P(x)\,dx} = e^{\ln|x-1|} = x - 1.$$

Therefore, multiplying both members of the standard form of the differential equation yields

$$y'(x - 1) + y = (x + 1)(x - 1)$$

$$\frac{d}{dx}[y(x - 1)] = (x + 1)(x - 1)$$

$$y(x - 1) = \int (x - 1)(x + 1)\,dx$$

$$= \int (x^2 - 1)\,dx = \frac{x^3}{3} - x + C_1$$

$$y = \frac{x^3 - 3x + C}{3(x - 1)}.$$

**15.** $y' + y\tan x = \sec x + \cos x$

The integrating factor is given by

$$e^{\int P\,dx} = e^{\int \tan x\,dx} = e^{\ln|\sec x|} = \sec x.$$

Therefore, multiplying both members of the differential equation in standard form yields

$$y'(\sec x) + (\sec x \tan x)y = \sec x(\sec x + \cos x)$$

$$\frac{d}{dx}[y\sec x] = \sec^2 x + 1$$

$$y\sec x = \int (\sec^2 x + 1)\,dx$$

$$= \tan x + x + C$$

$$y = \cos x(\tan x + x + C)$$

$$= \sin x + x\cos x + C\cos x.$$

Since $y = 1$ when $x = 0$, you have

$$1 = \sin 0 + 0\cos 0 + C\cos 0$$

$$1 = C.$$

Therefore, the particular solution is

$$y = \sin x + (x + 1)\cos x.$$

**21.** $y' + \left(\dfrac{1}{x}\right)y = xy^2.$

For the given Bernoulli equation we have $n = 2$, and $1 - n = -1$. If $z = y^{1-n} = y^{-1}$, then $z' = y^{-2}y'$. Multiplying both members of the differential equation by $-y^{-2}$ produces

$$y' + \left(\frac{1}{x}\right)y = xy^2$$

$$-y^{-2}y' - \left(\frac{1}{x}\right)y^{-1} = -x$$

$$z' - \left(\frac{1}{x}\right)z = -x.$$

This equation is linear in $z$. Using $P(x) = -1/x$ produces

$$-\int \frac{1}{x}\,dx = -\ln|x| \Rightarrow e^{-\ln|x|} = \frac{1}{x}$$

as the integrating factor. Multiplying the linear equation by this factor produces

$$z'\left(\frac{1}{x}\right) - \left(\frac{1}{x^2}\right)z = -1$$

$$\frac{d}{dx}\left[\frac{z}{x}\right] = -1$$

$$\frac{z}{x} = \int(-1)\,dx = -x + C$$

$$z = -x^2 + Cx$$

$$\frac{1}{y} = -x^2 + Cx$$

$$y = \frac{1}{Cx - x^2}.$$

**27.** $\dfrac{dy}{dx} + (\cot x)y = x$

(a) The graph of the direction field is shown in the figure.

(b) The integrating factor is

$$e^{\int P\,dx} = e^{\int \cot x\,dx} = e^{\ln \sin x} = \sin x.$$

Multiplying both sides of the differential equation by the integrating factor yields

$$\sin x\frac{dy}{dx} + \sin x(\cot x)y = x\sin x$$

$$\sin x\frac{dy}{dx} + y\cos x = x\sin x$$

$$\frac{d}{dx}[y\sin x] = x\sin x$$

$$y\sin x = \int x\sin x\,dx.$$

Using Integration by Parts, we have

$$y\sin x = \sin x - x\cos x + C.$$

We now find the particular solution passing through the point $(1, 1)$.

$$\sin 1 = \sin 1 - \cos 1 + C \Rightarrow C = \cos 1$$

—CONTINUED—

**27. —CONTINUED—**

Therefore,

$$y \sin x = \sin x - x \cos x + \cos 1 \text{ and}$$

$$y = 1 - x \cot x + \cos 1 \csc x.$$

We next find the particular solution passing through the point $(3, -1)$.

$$-\sin 3 = \sin 3 - 3 \cos 3 + C \Rightarrow C = 3 \cos 3 - 2 \sin 3$$

Therefore,

$$y \sin x = \sin x - x \cos x + 3 \cos 3 - 2 \sin 3 \text{ and}$$

$$y = 1 - x \cot x + (3 \cos 3 - 2 \sin 3) \csc x.$$

(c) The graphs of the particular solutions are shown in the figure.

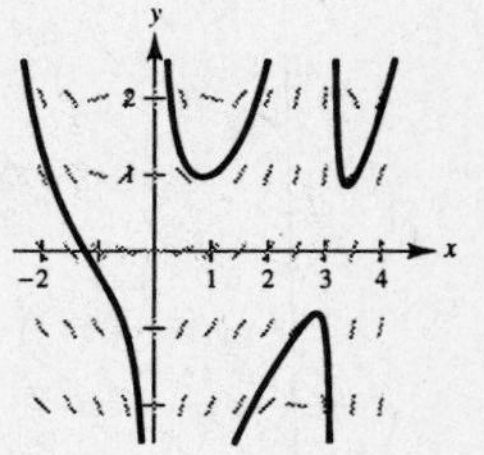

**37.** (a) Let $Q(t)$ be the amount of glucose in the blood stream at any time. The rate of change of $Q$ is given by

$$\frac{dQ}{dt} = \text{(rate administered)} - \text{(rate removed)}$$

$$\frac{dQ}{dt} = q - kQ$$

$$\frac{dQ}{dt} + kQ = q$$

where $t$ is time in minutes and $k$ is a constant of proportionality.

(b) To solve this linear equation begin by finding the integrating factor

$$e^{\int P(t)\,dt} = e^{\int k\,dt} = e^{kt}.$$

Therefore, the general solution has the form

$$Qe^{kt} = \int qe^{kt}\,dt = \frac{q}{k}e^{kt} + C$$

$$Q(t) = \frac{q}{k} + Ce^{-kt}.$$

Since $Q(0) = Q_0$, we have

$$Q_0 = \frac{q}{k} + C$$

$$Q_0 - \frac{q}{k} = C$$

$$Q(t) = \frac{q}{k} + \left(Q_0 - \frac{q}{k}\right)e^{-kt}.$$

(c) $\lim_{t\to\infty} Q(t) = \lim_{t\to\infty} \left[\frac{q}{k} + \left(Q_0 - \frac{q}{k}\right)e^{-kt}\right] = \frac{q}{k}$

**41.** (a) Using Exercise 39, we have $r_1 = 10$, $r_2 = 10$, $q_1 = 0$, and $v_0 = 200$. Thus,

$$\frac{dQ}{dt} + \frac{10Q}{200 + (0)t} = 0$$

$$\frac{dQ}{dt} = \frac{-10Q}{200} = -\frac{Q}{20}.$$

Separating variables, we have

$$\frac{dQ}{Q} = -\frac{dt}{20}$$

$$\ln|Q| = -\frac{t}{20} + C_1$$

$$Q(t) = e^{C_1 - (t/20)} = Ce^{-t/20}.$$

Since $Q(0) = 25$, we have

$$Q(t) = 25e^{-t/20}.$$

(b) When $Q = 15$, we have

$$15 = 25e^{-t/20}$$

$$\frac{3}{5} = e^{-t/20}$$

$$\ln 3 - \ln 5 = -\frac{t}{20}$$

$$t = 20(\ln 5 - \ln 3) \approx 10.2 \text{ min.}$$

(c) $\lim_{t\to\infty} Q(t) = \lim_{t\to\infty} Ce^{-t/20} = 0$

**49.** $\dfrac{dy}{dx} = \dfrac{e^{2x+y}}{e^{x-y}}$

Separating variables yields

$$e^{2x+y}\,dx = e^{x-y}\,dy$$
$$e^{2x}e^{y}\,dx = e^{x}e^{-y}\,dy$$
$$e^{x}\,dx = e^{-2y}\,dy$$
$$\int e^{x}\,dx = \int e^{-2y}\,dy$$
$$e^{x} = -\frac{1}{2}e^{-2y} + C_1$$
$$2e^{x} + e^{-2y} = C.$$

**55.** $2xy\,dx + (x^2 + \cos y)\,dy = 0$

Since

$$\frac{\partial M}{\partial y} = 2x = \frac{\partial N}{\partial x}$$

the equation is exact. Thus,

$$f(x, y) = \int 2xy\,dx + g(y) = x^2y + g(y)$$
$$f_y(x, y) = x^2 + g'(y) = \overbrace{x^2 + \cos y}^{N(x,\,y)}.$$

Therefore,

$$g(y) = \int \cos y\,dy = \sin y + C_1$$

and

$$f(x, y) = x^2y + \sin y + C_1.$$

Hence, the general solution is

$$x^2y + \sin y = C.$$

**61.** $(x^2y^4 - 1)\,dx + x^3y^3\,dy = 0$

Rewriting the differential equation we have

$$x^3y^3\frac{dy}{dx} + x^2y^4 = 1$$
$$\frac{dy}{dx} + \left(\frac{1}{x}\right)y = x^{-3}y^{-3}.$$

For this Bernoulli equation, $n = -3$ and use the substitution $z = y^{1-n} = y^4$ and $z' = 4y^3y'$. Multiplying the equation in standard form by $4y^3$ produces

$$4y^3y' + 4\left(\frac{1}{x}\right)y^4 = \frac{4}{x^3}$$
$$z' + \left(\frac{4}{x}\right)z = \frac{4}{x^3}.$$

This equation in linear in $z$. Using $P(x) = 4/x$ produces

$$\int P(x)\,dx = \int \frac{4}{x}\,dx = 4\ln|x| = \ln x^4$$

which implies that $e^{\ln x^4} = x^4$ is an integrating factor. Multiplying the linear equation by this factor produces

$$x^4z' + 4x^3z = 4x$$
$$\frac{d}{dx}[x^4z] = 4x$$
$$x^4z = \int 4x\,dx = 2x^2 + C$$
$$x^4y^4 = 2x^2 + C$$
$$x^4y^4 - 2x^2 = C.$$

## Section 15.3 Second-Order Homogeneous Linear Equations

**7.** $y'' - y' - 6y = 0$

The characteristic equation

$$m^2 - m - 6 = 0 \quad \text{or} \quad (m-3)(m+2) = 0$$

has two distinct real roots, $m_1 = 3$ and $m_2 = -2$. Thus the general solution is

$$y = C_1e^{m_1x} + C_2e^{m_2x} = C_1e^{3x} + C_2e^{-2x}.$$

**11.** $y'' + 6y' + 9y = 0$

The characteristic equation

$$m^2 + 6m + 9 = 0 \quad \text{or} \quad (m+3)^2 = 0$$

has two equal roots given by $m = -3$. Thus the general solution is

$$y = C_1e^{m_1x} + C_2xe^{m_1x} = C_1e^{-3x} + C_2xe^{-3x}$$
$$= (C_1 + C_2x)e^{-3x}.$$

**21.** $y'' - 3y' + y = 0$

The characteristic equation

$$m^2 - 3m + 1 = 0$$

has the distinct roots (using the quadratic formula)

$$m = \frac{3 \pm \sqrt{9-4}}{2} = \frac{3 \pm \sqrt{5}}{2}.$$

Thus, the general solution is

$$y = C_1e^{(3+\sqrt{5})x/2} + C_2e^{(3-\sqrt{5})x/2}.$$

**23.** $9y'' - 12y' + 11y = 0$

The characteristic equation

$$9m^2 - 12m + 11 = 0$$

has complex roots.

$$m = \frac{12 \pm \sqrt{144 - 4(9)(11)}}{2(9)}$$
$$= \frac{12 \pm \sqrt{-252}}{18} = \frac{12 \pm 6i\sqrt{7}}{18} = \frac{2}{3} \pm \frac{\sqrt{7}}{3}i$$

Thus, $\alpha = 2/3$ and $\beta = \sqrt{7}/3$ and the general solution is

$$y = C_1e^{2x/3}\cos\frac{\sqrt{7}x}{3} + C_2e^{2x/3}\sin\frac{\sqrt{7}x}{3}$$
$$= e^{2x/3}\left[C_1\cos\frac{\sqrt{7}x}{3} + C_2\sin\frac{\sqrt{7}x}{3}\right].$$

**33.** $y'' - y' - 30y = 0$

The characteristic equation

$$m^2 - m - 30 = 0 \quad \text{or} \quad (m-6)(m+5) = 0$$

has two distinct real roots, $m_1 = -5$ and $m_2 = 6$. Thus, the general solution is

$$y = C_1e^{m_1x} + C_2e^{m_2x} = C_1e^{-5x} + C_2e^{6x}.$$

Since $y(0) = 1$, we have

$$1 = C_1e^0 + C_2e^0 \quad \text{or} \quad C_1 + C_2 = 1.$$

Differentiating the general solution and using the initial condition $y'(0) = -4$, produces the following.

$$y' = -5C_1e^{-5x} + 6C_2e^{6x}$$
$$-4 = -5C_1e^0 + 6C_2e^0 \quad \text{or} \quad -5C_1 + 6C_2 = -4$$

Now solve the system of equations

$$C_1 + C_2 = 1$$
$$-5C_1 + 6C_2 = -4$$

to obtain $C_1 = \frac{10}{11}$ and $C_2 = \frac{1}{11}$. Hence, the particular solution is

$$y = \tfrac{10}{11}e^{-5x} + \tfrac{1}{11}e^{6x} = \tfrac{1}{11}(10e^{-5x} + e^{6x}).$$

**43.** By Hooke's Law, $32 = k(2/3)$ so that $k = 48$. Moreover, since the weight $w$ is given by $mg$, it follows that $m = w/g = \frac{32}{32} = 1$. Also the damping force is given by $-\frac{1}{8}(dy/dt)$. Thus the differential equation modeling the oscillations of the weight is

$$\frac{d^2y}{dt^2} = -\frac{1}{8}\left(\frac{dy}{dt}\right) - 48y$$

$$\frac{d^2y}{dt^2} + \frac{1}{8}\left(\frac{dy}{dt}\right) + 48y = 0.$$

The characteristic equation is

$$8m^2 + m + 384 = 0$$

with complex roots

$$m = -\frac{1}{16} \pm \frac{\sqrt{12{,}287}i}{16}.$$

Therefore, the general solution is

$$y(t) = e^{-t/16}\left(C_1 \cos\frac{\sqrt{12{,}287}t}{16} + C_2 \sin\frac{\sqrt{12{,}287}t}{16}\right).$$

Using the initial conditions, we have

$$y(0) = C_1 = \frac{1}{2}$$

$$y'(0) = e^{-t/16}\left[\left(-\frac{\sqrt{12{,}287}}{16}C_1 - \frac{C_2}{16}\right)\sin\frac{\sqrt{12{,}287}t}{16} + \left(\frac{12{,}287}{16}C_2 - \frac{C_1}{16}\right)\cos\frac{\sqrt{12{,}287}t}{16}\right]$$

$$y'(0) = \frac{\sqrt{12{,}287}}{16}C_2 - \frac{C_1}{16} = 0 \Rightarrow C_2 = \frac{\sqrt{12{,}287}}{24{,}574}.$$

and the particular solution

$$y(t) = \frac{e^{-t/16}}{2}\left[\cos\frac{\sqrt{12{,}287}t}{16} + \frac{\sqrt{12{,}287}}{12{,}287}\sin\frac{\sqrt{12{,}287}t}{16}\right].$$

**57.** $y_1 = e^{ax}\sin bx$ and $y_2 = e^{ax}\cos bx$.

$$W(y_1, y_2) = \begin{vmatrix} y_1 & y_2 \\ y_1' & y_2' \end{vmatrix}$$

$$= \begin{vmatrix} e^{ax}\sin bx & e^{ax}\cos bx \\ e^{ax}(b\cos bx + a\sin bx) & e^{ax}(a\cos bx - b\sin bx) \end{vmatrix}$$

$$= e^{2ax}\begin{vmatrix} \sin bx & \cos bx \\ b\cos bx + a\sin bx & a\cos bx - b\sin bx \end{vmatrix}$$

$$= e^{2ax}(a\sin bx\cos bx - b\sin^2 bx - b\cos^2 bx - a\sin bx\cos bx)$$

$$= -be^{2ax}$$

Since $W(y_1, y_2) \neq 0$, $y_1$ and $y_2$ are linearly independent.

## Section 15.4 Second-Order Nonhomogeneous Linear Equations

**7.** $y'' + y = x^3$

The characteristic equation $m^2 + 1 = 0$ has roots $m = \pm i$. Thus,

$$y_h = C_1 \cos x + C_2 \sin x.$$

Since $F(x) = x^3$, choose $y_p$ to be

$$y_p = A + Bx + Cx^2 + Dx^3.$$

Thus,

$$y_p' = B + 2Cx + 3Dx^2 \text{ and } y_p'' = 2C + 6Dx.$$

Substitution into the differential equation yields

$$y'' + y = x^3$$
$$(2C + 6Dx) + (A + Bx + Cx^2 + Dx^3) = x^3$$
$$(2C + A) + (6D + B)x + Cx^2 + Dx^3 = x^3.$$

Therefore, $2C + A = 0$, $6D + B = 0$, $C = 0$, and $D = 1$ from which it follows that $A = 0$ and $B = -6$. Thus, the general solution is

$$y = y_h + y_p = C_1 \cos x + C_2 \sin x - 6x + x^3.$$

Since $y(0) = 1$, $y'(0) = 0$, and $y' = -C_1 \sin x + C_2 \cos x - 6 + 3x^2$, we have

$$1 = C_1(1) + C_2(0) = C_1$$
$$0 = -C_1(0) + C_2(1) - 6 \quad \text{or} \quad C_2 = 6.$$

Finally, the particular solution is

$$y = \cos x + 6 \sin x - 6x + x^3.$$

**15.** $y'' + 9y = \sin 3x$

The characteristic equation $m^2 + 9 = 0$ has roots $m = \pm 3i$ and we have

$$y_h = C_1 \cos 3x + C_2 \sin 3x$$

Since $F(x) = \sin 3x$, consider

$$y_p = A \cos 3x + B \sin 3x$$

However, the terms of $y_p$ are *not* independent of those of $y_h$. Thus, multiply both terms by $x$ and obtain

$$y_p = Ax \cos 3x + Bx \sin 3x$$
$$y_p' = -3Ax \sin 3x + A \cos 3x + 3Bx \cos 3x + B \sin 3x$$
$$y_p'' = -6A \sin 3x + 6B \cos 3x - 9Ax \cos 3x - 9Bx \sin 3x$$

Substitution into the given differential equation and simplifying yields

$$y_p'' + 9y_p = -6A \sin 3x + 6B \cos 3x = \sin 3x.$$

Therefore, $A = -\frac{1}{6}$ and $B = 0$ and the general solution is

$$y = y_h + y_p = C_1 \cos 3x + C_2 \sin 3x - \frac{x}{6} \cos 3x$$
$$= \left(C_1 - \frac{x}{6}\right)\cos 3x + C_2 \sin 3x.$$

**25.** $y'' + 4y = \csc 2x$

The characteristic equation $m^2 + 4 = 0$, has solutions $m = \pm 2i$. Hence,

$$y_h = C_1 \cos 2x + C_2 \sin 2x.$$

Replacing $C_1$ and $C_2$ by $u_1$ and $u_2$ produces

$$y_p = u_1 \cos 2x + u_2 \sin 2x.$$

By the method of variation of parameters we obtain the following system of equations.

$$u_1' \cos 2x + u_2' \sin 2x = 0$$

$$u_1'(-2 \sin 2x) + u_2'(2 \cos 2x) = \csc 2x$$

Multiplying the first equation by $2 \sin 2x$ and the second by $\cos 2x$, and then adding the equations yields $u_2' = \frac{1}{2} \cot 2x$.

Substituting this result into the first equation, yields $u_1' = -\frac{1}{2}$. Integration yields

$$u_1 = \int -\frac{1}{2}\, dx = -\frac{x}{2}$$

$$u_2 = \int \frac{1}{2} \cot 2x\, dx = \frac{1}{4} \ln|\sin 2x|$$

and it follows that

$$\begin{aligned} y &= y_h + y_p \\ &= C_1 \cos 2x + C_2 \sin 2x - \frac{x}{2} \cos 2x + \frac{1}{4} \sin 2x \ln|\sin 2x| \\ &= \left(C_1 - \frac{x}{2}\right) \cos 2x + \left(C_2 + \frac{1}{4} \ln|\sin 2x|\right) \sin 2x. \end{aligned}$$

**31.** $\frac{24}{32} y'' + 48y = \frac{24}{32}(48 \sin 4t)$

$$y'' + 64y = 48 \sin 4t$$

The characteristic equation, $m^2 + 64 = 0$, has roots $m = \pm 8i$ and we have

$$y_h = C_1 \cos 8t + C_2 \sin 8t.$$

Since the derivatives of even order of the sine function is a sine function, let $y_p = A \sin 4t$. Therefore, $y_p' = 4A \cos 4t$ and $y_p'' = -16 A \sin 4t$. Substituting these results in the differential equation and simplifying yields

$$y_p'' + 64y_p = -16A \sin 4t + 64 A \sin 4t = 48A \sin 4t = 48 \sin 4t.$$

Thus, $A = 1$ and

$$y = y_h + y_p = C_1 \cos 8t + C_2 \sin 8t + \sin 4t$$

$$y' = y_h' + y_p' = -8C_1 \sin 8t + 8C_2 \cos 8t + 4 \cos 4t.$$

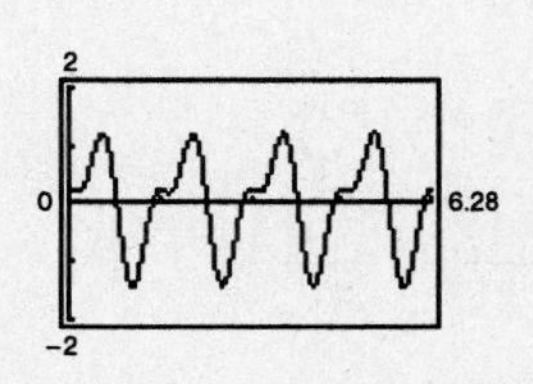


Since $y = \frac{1}{4}$ and $y' = 0$ when $t = 0$, we have

$$\frac{1}{4} = C_1$$

$$0 = 8C_2 + 4 \quad \text{or} \quad C_2 = -\frac{1}{2}.$$

Therefore,

$$y = \frac{1}{4} \cos 8t - \frac{1}{2} \sin 8t + \sin 4t.$$

## Section 15.5 Series Solutions of Differential Equations

**3.** $y'' - 9y = 0$

Assume $y = \sum_{n=0}^{\infty} a_n x^n$ is a solution. Then,

$$y' = \sum_{n=1}^{\infty} n a_n x^{n-1} \quad \text{and} \quad y'' = \sum_{n=2}^{\infty} n(n-1)a_n x^{n-2}.$$

Substituting these series in the differential equation yields

$$y'' - 9y = 0$$

$$\sum_{n=2}^{\infty} n(n-1)a_n x^{n-2} - 9\sum_{n=0}^{\infty} a_n x^n = 0$$

$$\sum_{n=2}^{\infty} n(n-1)a_n x^{n-2} = \sum_{n=0}^{\infty} 9a_n x^n.$$

The index of summation is changed by replacing $n$ by $n + 2$ in the left sum to insure that $x^n$ occurs in both sums. Thus,

$$\sum_{n=0}^{\infty} (n+2)(n+1)a_{n+2}x^n = \sum_{n=0}^{\infty} 9a_n x^n.$$

Equating coefficients yields

$$(n+2)(n+1)a_{n+2} = 9a_n$$

from which we obtain the recursion formula

$$a_{n+2} = \frac{9a_n}{(n+2)(n+1)} \quad (n \geq 0).$$

Thus the coefficients of the series solution are

$$a_2 = \frac{9}{2}a_0 = \frac{3^2}{2}a_0 \qquad a_3 = \frac{9}{6}a_1 = \frac{3^2}{3 \cdot 2}a_1$$

$$a_4 = \frac{9}{12}a_2 = \frac{3^4}{4 \cdot 3 \cdot 2}a_0 \qquad a_5 = \frac{9}{20}a_3 = \frac{3^4}{5 \cdot 4 \cdot 3 \cdot 2}a_1$$

$$\vdots \qquad\qquad \vdots$$

$$a_{2k} = \frac{(3)^{2k}}{(2k)!}a_0 \qquad a_{2k+1} = \frac{(3)^{2k}}{(2k+1)!}a_1$$

Thus, we can represent the general solution as the sum of two power series—one for the even-powered terms with coefficients in terms of $a_0$ and one for the odd-powered terms with coefficients in terms of $a_1$.

$$y = \left(a_0x^0 + \frac{3^2}{2!}a_0x^2 + \frac{3^4}{4!}a_0x^4 + \cdots\right) + \left(a_1x + \frac{3^2}{3!}a_1x^3 + \frac{3^4}{5!}a_1x^5 + \cdots\right)$$

$$= a_0\left[1 + \frac{(3x)^2}{2!} + \frac{(3x)^4}{4!} + \cdots\right] + \frac{a_1}{3}\left[3x + \frac{(3x)^3}{3!} + \frac{(3x)^5}{5!} + \cdots\right]$$

$$= a_0\sum_{k=0}^{\infty}\frac{(3x)^{2k}}{(2k)!} + \frac{a_1}{3}\sum_{k=0}^{\infty}\frac{(3x)^{2k+1}}{(2k+1)!}$$

Observe that $y'' - 9y = 0$ is a second-order homogeneous differential equation with characteristic equation $m^2 - 9 = 0$. Thus, the general solution is

$$y = C_1e^{3x} + C_2e^{-3x}.$$

**—CONTINUED—**

**3. —CONTINUED—**

The following reconciles this form of the solution with the series solution given above.

$$y = C_1e^{3x} + C_2e^{-3x}$$

$$= C_1\sum_{n=0}^{\infty}\frac{(3x)^n}{n!} + C_2\sum_{n=0}^{\infty}\frac{(-3x)^n}{n!}$$

$$= C_1\left[1 + (3x) + \frac{(3x)^2}{2!} + \cdots\right] + C_2\left[1 + (-3x) + \frac{(-3x)^2}{2!} + \cdots\right]$$

$$= C_1\left[1 + \frac{(3x)^2}{2!} + \frac{(3x)^4}{4!} + \cdots\right] + C_1\left[(3x) + \frac{(3x)^3}{3!} + \frac{(3x)^5}{5!} + \cdots\right]$$

$$+ C_2\left[1 + \frac{(-3x)^2}{2!} + \frac{(-3x)^4}{4!} + \cdots\right] + C_2\left[(-3x) + \frac{(-3x)^3}{3!} + \frac{(-3x)^5}{5!} + \cdots\right]$$

$$= (C_1 + C_2)\left[1 + \frac{(3x)^2}{2!} + \frac{(3x)^4}{4!} + \cdots\right] + (C_1 - C_2)\left[(3x) + \frac{(3x)^3}{3!} + \cdots\right]$$

$$= a_0\sum_{k=0}^{\infty}\frac{(3x)^{2k}}{(2k)!} + \frac{a_1}{3}\sum_{k=0}^{\infty}\frac{(3x)^{2k+1}}{(2k+1)!}$$

where $C_1 + C_2 = a_0$ and $C_1 - C_2 = a_1/3$.

**9.** $y'' - xy' = 0$

Assume $y = \sum_{n=0}^{\infty} a_nx^n$ is a solution. Then $y' = \sum_{n=1}^{\infty} na_nx^{n-1}$ and $y'' = \sum_{n=2}^{\infty} n(n-1)a_nx^{n-2}$.

Thus, the equation $y'' - xy' = 0$ is written as

$$\sum_{n=2}^{\infty} n(n-1)a_nx^{n-2} - \sum_{n=0}^{\infty} na_nx^n = 0$$

$$\sum_{n=2}^{\infty} n(n-1)a_nx^{n-2} = \sum_{n=0}^{\infty} na_nx^n.$$

The index of summation is changed by replacing $n$ by $n + 2$ in the left sum to insure that $x^n$ occurs in both sums. Thus,

$$\sum_{n=0}^{\infty}(n+2)(n+1)a_{n+2}x^n = \sum_{n=0}^{\infty} na_nx^n.$$

Equating coefficients, you have

$$(n+2)(n+1)a_{n+2} = na_n$$

$$a_{n+2} = \frac{na_n}{(n+2)(n+1)}. \qquad \text{(Recursion Formula)}$$

Thus, the coefficients of the series solution are

$$a_2 = \frac{0}{2}a_0 = 0 \qquad a_3 = \frac{1}{3\cdot 2}a_1$$

$$a_4 = \frac{2}{12}a_2 = 0 \qquad a_5 = \frac{3}{5\cdot 4}a_3 = \frac{1\cdot 3}{2\cdot 3\cdot 4\cdot 5}a_1$$

$$a_7 = \frac{5}{7\cdot 6}a_5 = \frac{1\cdot 3\cdot 5}{2\cdot 3\cdot 4\cdot 5\cdot 6\cdot 7}a_1$$

$$\vdots \qquad \vdots$$

$$a_{2k} = 0 \qquad a_{2k+1} = \frac{1\cdot 3\cdot 5\cdots(2k-1)}{1\cdot 2\cdot 3\cdot 4\cdot 5\cdots(2k+1)}a_1 = \frac{a_1}{2\cdot 4\cdot 6\ldots(2k)(2k+1)}$$

$$= \frac{a_1}{2^k(1\cdot 2\cdot 3\ldots k)(2k+1)} = \frac{a_1}{2^kk!(2k+1)}.$$

**—CONTINUED—**

**9. —CONTINUED—**

Since the coefficients of the terms with even powers of $x$ are all zero, the solution is

$$y = a_1 \sum_{k=0}^{\infty} \frac{x^{2k+1}}{2^k k!(2k+1)}.$$

We can find the interval of convergence by using the Ratio Test with $u_k = \dfrac{x^{2k+1}}{2^k k!(2k+1)}$.

$$\lim_{k\to\infty} \left|\frac{u_{k+1}}{u_k}\right| = \lim_{k\to\infty} \left|\frac{x^{2k+3}}{2^{k+1}(k+1)!(2k+3)} \cdot \frac{2^k k!(2k+1)}{x^{2k+1}}\right|$$

$$= \lim_{k\to\infty} \left|\frac{(2k+1)x^2}{2(k+1)(2k+3)}\right| = 0$$

for any value of $x$. Therefore, the interval of convergence is $(-\infty, \infty)$.

**17.** $y'' - 2xy = 0,\ y(0) = 1,\ y'(0) = -3$

Taylor's Theorem for $c = 0$ is $y = y(0) + y'(0)x + \dfrac{y''(0)}{2!}x^2 + \dfrac{y'''(0)}{3!}x^3 + \cdots$.

Since $y'' = 2xy$, $y(0) = 1$ and $y'(0) = -3$, we have

$$\begin{aligned}
& & y(0) &= 1\\
& & y'(0) &= -3\\
y'' &= 2xy & y''(0) &= 0\\
y''' &= 2xy' + 2y & y'''(0) &= 2\\
y^{(4)} &= 2xy'' + 4y' & y^{(4)}(0) &= -12\\
y^{(5)} &= 2xy''' + 6y'' & y^{(5)}(0) &= 0\\
y^{(6)} &= 2xy^{(4)} + 8y''' & y^{(6)}(0) &= 16\\
y^{(7)} &= 2xy^{(5)} + 10y^{(4)} & y^{(7)}(0) &= -120.
\end{aligned}$$

Therefore, the first six terms of the Taylor series are

$$y = 1 - 3x + 0x^2 + \frac{2}{3!}x^3 - \frac{12}{4!}x^4 + 0x^5 + \frac{16}{6!}x^6 - \frac{120}{7!}x^7 + \cdots$$

$$= 1 - 3x + \frac{1}{3}x^3 - \frac{1}{2}x^4 + \frac{1}{45}x^6 - \frac{1}{42}x^7 + \cdots.$$

Finally, at $x = \frac{1}{4}$ we have

$$y = 1 - \frac{3}{4} + \frac{1}{3 \cdot 4^3} - \frac{1}{2 \cdot 4^4} + \frac{1}{45 \cdot 4^6} + \frac{1}{42 \cdot 4^7} \approx 0.253.$$

**19.** $y = \displaystyle\sum_{n=0}^{\infty} \frac{x^n}{n!}$

Since the solution to a differential equation is unique, you only need to show that both the series and the exponential function are solutions to the given differential equation. Beginning with the series

$$y = \sum_{n=0}^{\infty} \frac{x^n}{n!}$$

$$y' = \sum_{n=1}^{\infty} \frac{nx^{n-1}}{n!} = \sum_{n=1}^{\infty} \frac{x^{n-1}}{(n-1)!} = \sum_{n=0}^{\infty} \frac{x^n}{n!} = y.$$

Therefore, the function represented by the series is a solution of the differential equation $y' - y = 0$. Since $y = e^x = y'$, the exponential function is also a solution to the differential equation.

# Review Exercises for Chapter 15

**3.** $y'' + 3y' - 10 = 0$

Since the function $y$ has only one independent variable, the equation is called an ordinary differential equation. The highest-order derivative in the equation is a second derivative. Therefore, the differential equation is second order.

**15.** $\dfrac{dy}{dx} - \dfrac{y}{x} = \dfrac{x}{y}$

The given equation

$$\frac{dy}{dx} - \left(\frac{1}{x}\right)y = xy^{-1}$$

is a Bernoulli equation with $n = -1$, and $1 - n = 2$. Let $z = y^{1-n} = y^2$, then $z' = 2yy'$. Multiplying the original differential equation by $2y$ produces

$$2yy' - 2\left(\frac{1}{x}\right)y^2 = 2x$$

$$z' - \left(\frac{2}{x}\right)z = 2x.$$

This equation is linear in $z$. Using $P(x) = -2/x$ yields

$$\int P(x)\,dx = \int \frac{-2}{x}\,dx = -2\ln x = \ln x^{-2}$$

which implies that $e^{\ln x^{-2}} = x^{-2}$ is an integrating factor. Multiplying the linear equation by this factor produces

$$\left(\frac{1}{x^2}\right)z' - \left(\frac{2}{x^3}\right)z = \frac{2}{x}$$

$$\frac{d}{dx}\left[\frac{1}{x^2}(z)\right] = \frac{2}{x}$$

$$z\left(\frac{1}{x^2}\right) = \int \frac{2}{x}\,dx = 2\ln|x| + C$$

$$y^2 = 2x^2\ln|x| + Cx^2 = x^2\ln x^2 + Cx^2.$$

**19.** $(2x - 2y^3 + y)\,dx + (x - 6xy^2)\,dy = 0$

Since

$$\frac{\partial M}{\partial y} = 1 - 6y^2 = \frac{\partial N}{\partial x}$$

the equation is exact. Therefore, there exists a function $f$ such that $f_x(x, y) = 2x - 2y^3 + y$ and $f_y(x, y) = x - 6xy^2$. Integration yields

$$f(x, y) = \int (2x - 2y^3 + y)\,dx = x^2 - 2xy^3 + xy + g(y)$$

$$f(x, y) = \int (x - 6xy^2)\,dy = xy - 2xy^3 + h(x)$$

Reconciling these two versions of $f$, yields $h(x) = x^2$, $g(y) = 0$ and $f(x, y) = x^2 - 2xy^3 + xy$. Therefore, the general solution $f(x, y) = C$ is

$$x^2 - 2xy^3 + xy = C.$$

**29.** $(1 + x^2)\,dy = (1 + y^2)\,dx$.

Since the variables are separable, we have

$$\frac{dy}{1 + y^2} = \frac{dx}{1 + x^2}$$

$$\int \frac{dy}{1 + y^2} = \int \frac{dx}{1 + x^2}$$

$$\arctan y = \arctan x + C_1.$$

Now taking the tangent of this equation and using the identity for $\tan(A - B)$, yields

$$\tan(\arctan y - \arctan x) = \tan C_1$$

$$\frac{\tan \arctan y - \tan \arctan x}{1 + \tan(\arctan y)\tan(\arctan x)} = C$$

$$\frac{y - x}{1 + xy} = C$$

as the general solution.

**35.**

$$x\,dy = (x + y + 2)\,dx$$

$$x\,dy = y\,dx + (x + 2)\,dx$$

$$dy = \left(\frac{1}{x}\right)y\,dx + \frac{x + 2}{x}\,dx$$

$$\frac{dy}{dx} - \left(\frac{1}{x}\right)y = \left(1 + \frac{2}{x}\right)$$

The integrating factor is

$$e^{\int P\,dx} = e^{\int(-1/x)\,dx} = e^{-\ln x} = e^{\ln(1/x)} = \frac{1}{x}.$$

Multiplying both sides of the differential equation by $1/x$ yields

$$\left(\frac{1}{x}\right)\frac{dy}{dx} - \left(\frac{1}{x^2}\right)y = \left(\frac{1}{x} + \frac{2}{x^2}\right)$$

$$\frac{d}{dx}\left[y\left(\frac{1}{x}\right)\right] = \left(\frac{1}{x} + \frac{2}{x^2}\right)$$

$$y\left(\frac{1}{x}\right) = \int\left(\frac{1}{x} + \frac{2}{x^2}\right)dx$$

$$y\left(\frac{1}{x}\right) = \ln|x| - \frac{2}{x} + C$$

$$y = x\ln|x| - 2 + Cx.$$

Since $y = 10$ when $x = 1$, we have

$$10 = 1\ln 1 - 2 + C \Rightarrow C = 12.$$

Therefore, the particular solution is

$$y = x\ln|x| - 2 + 12x.$$

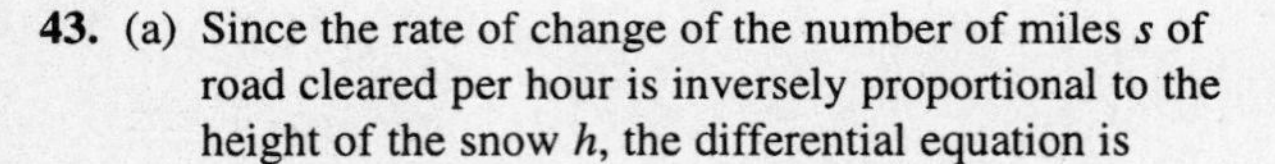

**43.** (a) Since the rate of change of the number of miles $s$ of road cleared per hour is inversely proportional to the height of the snow $h$, the differential equation is

$$\frac{ds}{dh} = \frac{k}{h}.$$

This is a separable differential equation. Therefore,

$$\frac{ds}{dh} = \frac{k}{h}$$

$$ds = k\frac{1}{h}\,dh$$

$$\int ds = k\int \frac{1}{h}\,dh$$

$$s = k\ln h + C.$$

(b) To find the particular solution use the fact that $s = 25$ when $h = 2$ and $s = 12$ when $h = 10$ to obtain the following system of equations:

$$25 = k\ln 2 + C$$

$$12 = k\ln 10 + C$$

Subtracting the second equation from the first yields

$$13 = k(\ln 2 - \ln 10) = k\ln\frac{1}{5} = -k\ln 5.$$

Thus, $k = -13/\ln 5$. Substituting this expression for $k$ into the first equation of the system produces

$$25 = -\frac{13}{\ln 5}\ln 2 + C \Rightarrow C = 25 + \frac{13}{\ln 5}\ln 2.$$

Substituting the results for $k$ and $C$ into the general solution of the differential equation yields the particular solution

$$s = -\frac{13}{\ln 5}\ln h + 25 + \frac{13}{\ln 5}\ln 2$$

$$= 25 - \frac{13}{\ln 5}(\ln h - \ln 2)$$

$$= 25 - \frac{13\ln(h/2)}{\ln 5}, \quad 2 \le h \le 15.$$

**49.** $\dfrac{dA}{dt} = rA - P$

$$\frac{dA}{dt} - rA = -P$$

The integrating factor is $e^{\int -r\,dt} = e^{-rt}$. Multiplying both sides of the differential equation by the integrating factor yields

$$e^{-rt}\frac{dA}{dt} - rAe^{-rt} = Pe^{-rt}$$

$$\frac{d}{dt}[Ae^{-rt}] = -Pe^{-rt}$$

$$Ae^{-rt} = \int -Pe^{-rt}\,dt$$

$$Ae^{-rt} = \frac{P}{r}e^{-rt} + C$$

$$A = \frac{P}{r} + Ce^{rt}.$$

If $A = A_0$ when $t = 0$, then

$$A_0 = \frac{P}{r} + C \Rightarrow C = A_0 - \frac{P}{r}$$

and the particular solution is

$$A = \frac{P}{r} + \left(A_0 - \frac{P}{r}\right)e^{rt}.$$

(a) $A_0 = \$500{,}000$, $P = \$40{,}000$, $r = 0.10$

$$A = \frac{40{,}000}{0.10} + \left(500{,}000 - \frac{40{,}000}{0.10}\right)e^{0.10t}$$

$$= 100{,}000(4 + e^{0.10t})$$

The balance continues to grow.

(b) $A_0 = \$500{,}000$, $P = \$50{,}000$, $r = 0.10$

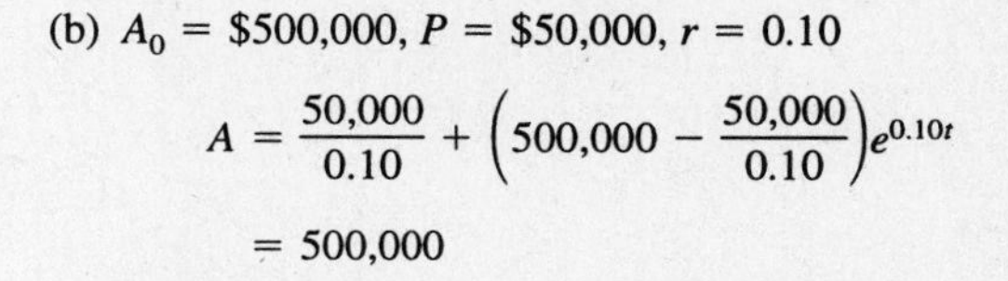

$$A = \frac{50{,}000}{0.10} + \left(500{,}000 - \frac{50{,}000}{0.10}\right)e^{0.10t}$$

$$= 500{,}000$$

The balance remains constant at \$500,000

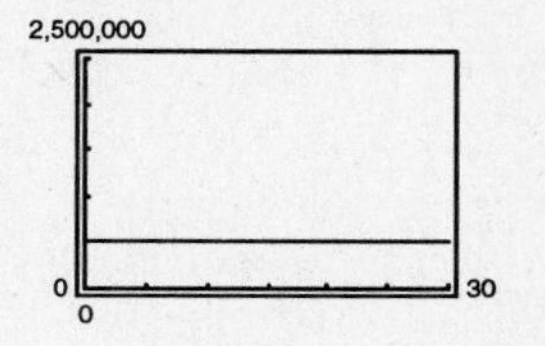


(c) $A_0 = \$500{,}000$, $P = \$60{,}000$, $r = 0.10$

$$A = \frac{60{,}000}{0.10} + \left(500{,}000 - \frac{60{,}000}{0.10}\right)e^{0.10t}$$

$$= 100{,}000(6 - e^{0.10t})$$

The balance decreases and is depleted when $6 - e^{0.10t} = 0$ or $t = \dfrac{\ln 6}{0.10} \approx 17.9$ years.

**59.** $y'' - 2y' + y = 2xe^x$

The characteristic equation $m^2 - 2m + 1 = 0$ has equal roots given by $m = 1$. Thus the general solution is

$$y_h = C_1e^x + C_2xe^x.$$

Since $F(x) = 2xe^x$, use the Method of Undetermined Coefficients with

$$y_p = (A + Bx)e^x = Ae^x + Bxe^x.$$

This function is not independent of $y_h$ since it already contains both of these terms. Therefore, multiply by $x^2$ to obtain

$$y_p = Ax^2e^x + Bx^3e^x = e^x(Ax^2 + Bx^3)$$
$$y_p' = 2Axe^x + (3B + A)x^2e^x + Bx^3e^x$$
$$y_p'' = 2Ae^x + (4A + 6B)xe^x + (A + 6B)x^2e^2 + Bx^3e^x.$$

Substituting into the given differential equation, yields

$$2Ae^x + (4A + 6B)xe^x + (A + 6B)x^2e^x + Bx^3e^x - 4Axe^x - 2(3B + A)x^2e^x - 2Bx^3e^x + Ax^2e^x + Bx^3e^x = 2xe^x$$

or

$$2Ae^x + 6Bxe^x = 2xe^x.$$

Equating coefficients, yields $A = 0$ and $6B = 2$, or $B = \frac{1}{3}$. Hence, $y_p = \frac{1}{3}x^3e^x$, and the general solution is

$$\begin{aligned} y &= y_h + y_p \\ &= C_1e^x + C_2xe^x + \frac{1}{3}x^3e^x \\ &= \left(C_1 + C_2x + \frac{1}{3}x^3\right)e^x. \end{aligned}$$

# Appendix F

**7.** $13i - (14 - 7i) = 13i - 14 + 7i = -14 + 20i$

**15.** $(1 + i)(3 - 2i) = 3 - 2i + 3i - 2i^2$
$= 3 + i - 2(-1) = 5 + i$

**23.** $(2 + 3i)^2 + (2 - 3i)^2 = [4 + 2(6i) + 9i^2] + [4 - 2(6i) + 9i^2]$
$= 4 + 12i + 9(-1) + 4 - 12i + 9(-1) = -10$

**27.** Complex number: $-2 - \sqrt{5}i$
Complex conjugate: $-2 + \sqrt{5}i$

$$\left(-2 - \sqrt{5}i\right)\left(-2 + \sqrt{5}i\right) = (-2)^2 - \left(\sqrt{5}i\right)^2$$
$$= 4 - 5i^2$$
$$= 4 - 5(-1) = 9$$

**35.** $$\frac{4}{4 - 5i} = \frac{4}{4 - 5i}\left(\frac{4 + 5i}{4 + 5i}\right)$$
$$= \frac{16 + 20i}{4^2 - (5i)^2}$$
$$= \frac{16 + 20i}{16 - 25(-1)}$$
$$= \frac{16 + 20i}{41} = \frac{16}{41} + \frac{20}{41}i$$

**43.** $$\frac{2}{1 + i} - \frac{3}{1 - i} = \frac{2}{1 + i}\left(\frac{1 - i}{1 - i}\right) - \frac{3}{1 - i}\left(\frac{1 + i}{1 + i}\right)$$
$$= \frac{2 - 2i}{1 - i^2} - \frac{3 + 3i}{1 - i^2}$$
$$= \frac{2 - 2i}{2} - \frac{3 + 3i}{2}$$
$$= 1 - i - \frac{3}{2} - \frac{3}{2}i = -\frac{1}{2} - \frac{5}{2}i$$

**49.** $4x^2 + 16x + 17 = 0$

Using the Quadratic Formula we have

$$x = \frac{-b \pm \sqrt{b^2 - 4ac}}{2a}$$
$$= \frac{-16 \pm \sqrt{16^2 - 4(4)(17)}}{2(4)}$$
$$= \frac{-16 \pm 4\sqrt{16 - 17}}{8} = \frac{-16 \pm 4i}{8} = -2 \pm \frac{1}{2}i.$$

**59.** $$\left(\sqrt{-75}\right)^3 = \left(\sqrt{3(5^2)(-1)}\right)^3$$
$$= \left(5\sqrt{3}i\right)^3$$
$$= 125\left(3\sqrt{3}\right)i^3 = 375\sqrt{3}(-i) = -375\sqrt{3}i$$

**69.** The graphical representation of $3 - 3i$ is shown in the figure. Its absolute value is

$$r = |3 - 3i| = \sqrt{3^2 + (-3)^2} = \sqrt{2(3^2)} = 3\sqrt{2}$$

and the tangent of the angle $\theta$ is given by

$$\tan\theta = \frac{b}{a} = \frac{-3}{3} = -1.$$

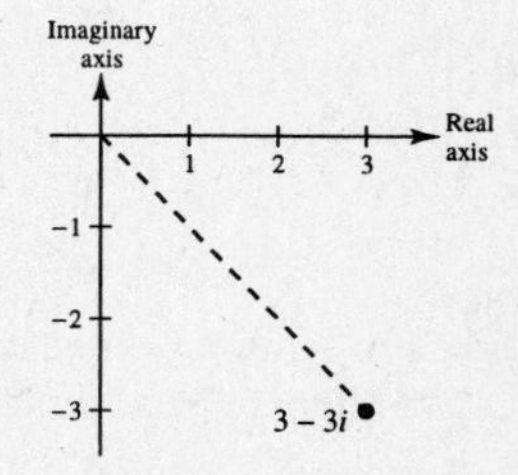


Because $\tan(\pi/4) = 1$ and $3 - 3i$ lies in Quadrant IV, we have $\theta = 2\pi - \pi/4 = 7\pi/4$. Thus the trigonometric form is

$$3 - 3i = r(\cos\theta + i\sin\theta) = 3\sqrt{2}\left(\cos\frac{7\pi}{4} + i\sin\frac{7\pi}{4}\right).$$

**79.** The graphical representation of $\frac{3}{2}(\cos 300° + i \sin 300°)$ is shown in the figure. Because $\cos 300° = \cos 60° = 1/2$ and $\sin 300° = -\sin 60° = -\sqrt{3}/2$, we can write

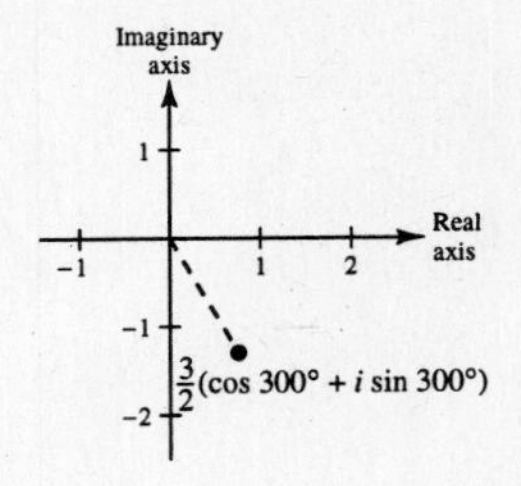


$$\frac{3}{2}(\cos 300° + i \sin 300°) = \frac{3}{2}\left(\frac{1}{2} - \frac{\sqrt{3}}{2}i\right)$$
$$= \frac{3}{4} - \frac{3\sqrt{3}}{4}i.$$

**89.** First convert to trigonometric form.

$$-1 + i = \sqrt{2}\left(\cos\frac{3\pi}{4} + i \sin\frac{3\pi}{4}\right)$$

Then, by DeMoivre's Theorem, we have

$$(-1 + i)^{10} = \left[\sqrt{2}\left(\cos\frac{3\pi}{4} + i \sin\frac{3\pi}{4}\right)\right]^{10}$$
$$= (2^{1/2})^{10}\left[\cos\left(10 \cdot \frac{3\pi}{4}\right) + i \sin\left(10 \cdot \frac{3\pi}{4}\right)\right]$$
$$= 32\left(\cos\frac{15\pi}{2} + i \sin\frac{15\pi}{2}\right)$$
$$= 32\left(\cos\frac{3\pi}{2} + i \sin\frac{3\pi}{2}\right) = 32(0 - i) = -32i.$$

**99.** (a) First convert to trigonometric form.

$$-\frac{125}{2}(1 + \sqrt{3}i) = 125\left(\cos\frac{4\pi}{3} + i \sin\frac{4\pi}{3}\right)$$

By the formula for $n$th roots, the cube roots have the form

$$\sqrt[3]{125}\left[\cos\frac{4\pi/3 + (2\pi)k}{3} + i \sin\frac{4\pi/3 + (2\pi)k}{3}\right].$$

Finally, for $k = 0$, 1, and 2, we obtain the following roots.

$$5\left(\cos\frac{4\pi}{9} + i \sin\frac{4\pi}{9}\right),\ 5\left(\cos\frac{10\pi}{9} + i \sin\frac{10\pi}{9}\right),$$
$$5\left(\cos\frac{16\pi}{9} + i \sin\frac{16\pi}{9}\right)$$

(b) The graphical representation of the roots is shown in the figure.

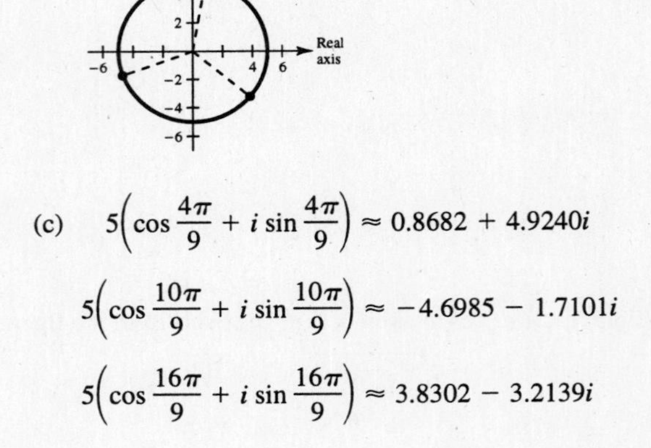


(c) $5\left(\cos\frac{4\pi}{9} + i \sin\frac{4\pi}{9}\right) \approx 0.8682 + 4.9240i$

$5\left(\cos\frac{10\pi}{9} + i \sin\frac{10\pi}{9}\right) \approx -4.6985 - 1.7101i$

$5\left(\cos\frac{16\pi}{9} + i \sin\frac{16\pi}{9}\right) \approx 3.8302 - 3.2139i$

**103.** The solutions of the equation $x^5 + 243 = 0$ are the five fifth roots of $-243$. First convert to trigonometric form

$$-243 = 243(\cos \pi + i \sin \pi).$$

By the formula of $n$th roots, the fifth roots have the form

$$\sqrt[5]{243}\left[\cos\frac{\pi + (2\pi)k}{5} + i \sin\frac{\pi + (2\pi)k}{5}\right].$$

Finally, for $k = 0$, 1, 2, 3, 4, and 5, we obtain the following roots.

$$3\left(\cos\frac{\pi}{5} + i \sin\frac{\pi}{5}\right),\ 3\left(\cos\frac{3\pi}{5} + i \sin\frac{3\pi}{5}\right),\ 3(\cos \pi + i \sin \pi),\ 3\left(\cos\frac{7\pi}{5} + i \sin\frac{7\pi}{5}\right),\ 3\left(\cos\frac{9\pi}{5} + i \sin\frac{9\pi}{5}\right)$$

The graphical representation of the roots is shown in the figure.